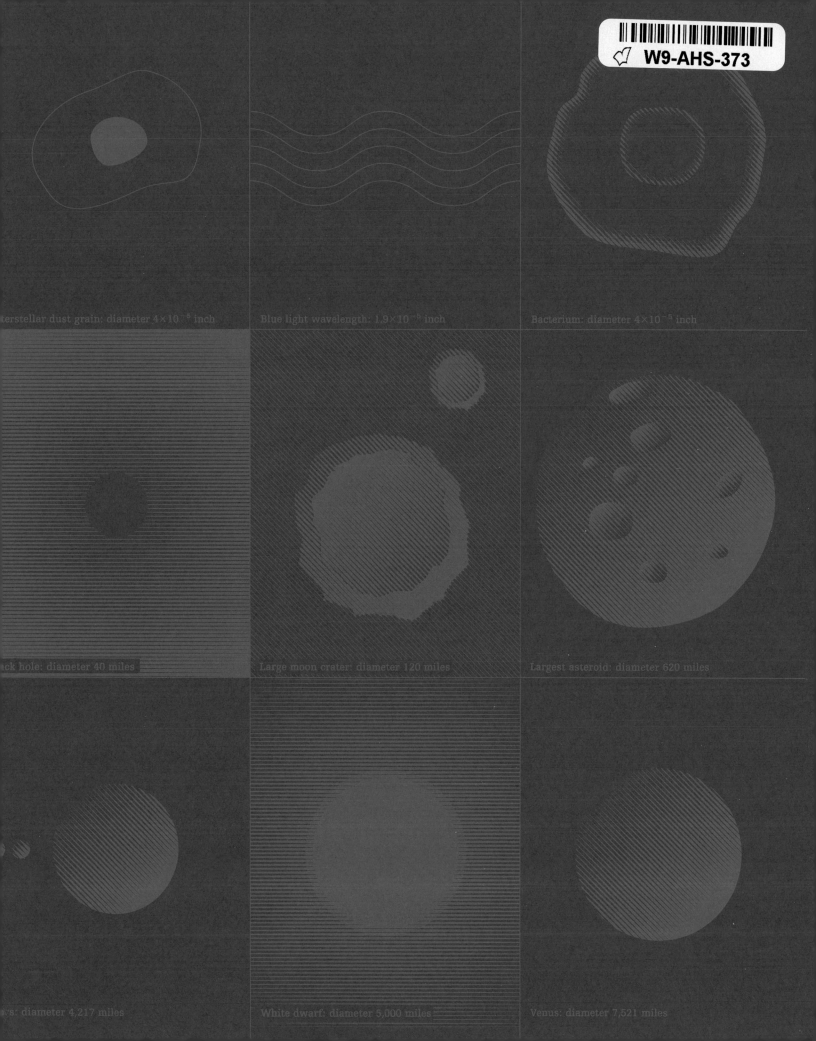

Interstellar dust grain: diameter 4×10^{-6} inch

Blue light wavelength: 1.9×10^{-5} inch

Bacterium: diameter 4×10^{-5} inch

Black hole: diameter 40 miles

Large moon crater: diameter 120 miles

Largest asteroid: diameter 620 miles

Mars: diameter 4,217 miles

White dwarf: diameter 5,000 miles

Venus: diameter 7,521 miles

THE COSMOS

Teeming stars in the southern sky suggest the unknowable immensity of the cosmos.

Subatomic collisions staged in a bubble chamber may offer clues to the origin of matter.

TIME LIFE ®

Other Publications:
THE THIRD REICH
THE TIME-LIFE GARDENER'S GUIDE
MYSTERIES OF THE UNKNOWN
TIME FRAME
FIX IT YOURSELF
FITNESS, HEALTH & NUTRITION
SUCCESSFUL PARENTING
HEALTHY HOME COOKING
UNDERSTANDING COMPUTERS
LIBRARY OF NATIONS
THE ENCHANTED WORLD
THE KODAK LIBRARY OF CREATIVE PHOTOGRAPHY
GREAT MEALS IN MINUTES
THE CIVIL WAR
PLANET EARTH
COLLECTOR'S LIBRARY OF THE CIVIL WAR
THE EPIC OF FLIGHT
THE GOOD COOK
WORLD WAR II
HOME REPAIR AND IMPROVEMENT
THE OLD WEST

This volume is one of a series that
examines the universe in all its aspects,
from its beginnings in the Big Bang to the
promise of space exploration.

VOYAGE THROUGH THE UNIVERSE

THE COSMOS

BY THE EDITORS OF TIME-LIFE BOOKS
ALEXANDRIA, VIRGINIA

CONTENTS

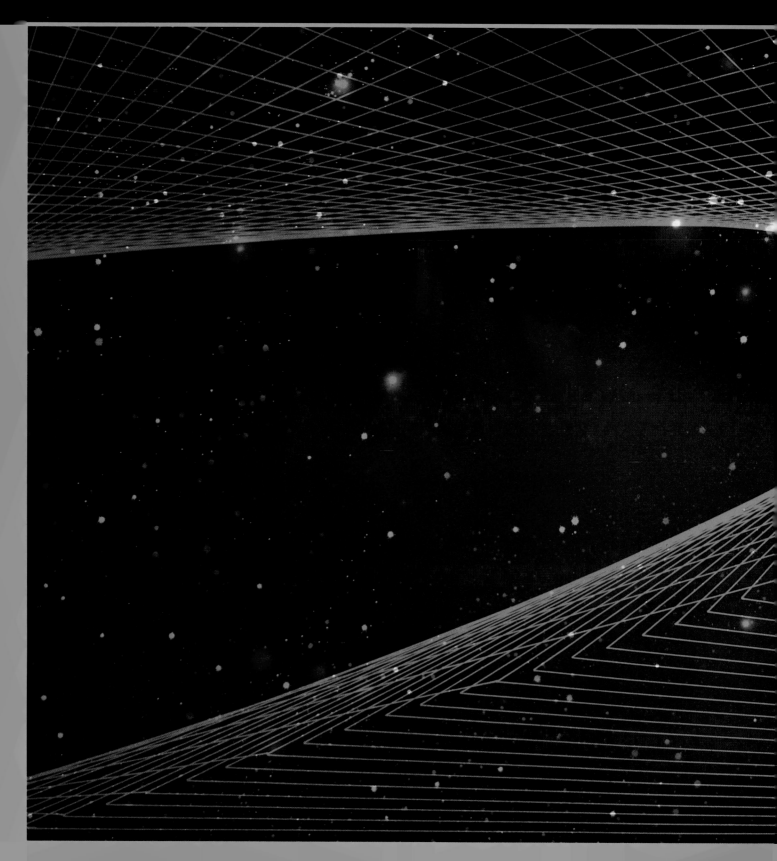

space and time, a curved universe is symbolized by this composite photograph. In predicting the warped geometry of such a cosmos, Albert Einstein's theory of relativity changed the course of science.

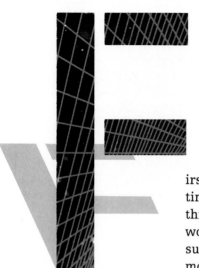

irst there is nothing—not time, not space, not even emptiness, since there is no space to be empty. Then, from this void, this utter nothingness so complete that no word can make it imaginable, springs . . . a universe, suddenly there but far smaller than the smallest dust mote. The seed for everything that will ever be, it contains all of creation. For now, however, chaos reigns; the universe is so tiny, hot, and dense that none of the familiar laws of physics hold. The dimensions of space and time are torn and snarled by discontinuities. The concepts *here* and *there* have no meaning, nor do *now* and *then.* There is no matter, no force such as gravity or electromagnetism, only a node of pure energy.

No sooner has it appeared than the pinpoint cosmos starts to expand, cooling as it grows. Within its still-infinitesimal confines, jumbled space begins to untangle. Time settles down to run from past to future, and order emerges. Quickly, within the merest sliver of an instant, the universe cools enough to allow gravity to congeal from the undifferentiated energy. Its force acts to slow the rate of expansion. Pairs of particles that can exist only in the extreme conditions of this era flicker spontaneously into existence. Closely packed and flashing through space, the bits of matter smash together in a mayhem of collisions. Often they annihilate each other, disappearing in a burst of energy. Sometimes they strike off a shower of new particles, all hastening to their own violent ends.

Soon, rapid cooling caused by the enlargement has spawned an environment so bizarre that the gravitational force is turned inside out. Instead of braking the expansion of the universe as it normally would, gravity causes the swelling to accelerate explosively. At once the cosmos erupts from subatomic proportions to the size of a grapefruit. New particles spring to life, swiftly grow more massive, and then decay into other particles, the stuff of atoms. Before the universe is a second old, it has grown as big as the Solar System, yet it is more dense than water and far hotter than the central furnace of a star. The searing contents of this crucible have become familiar forms of matter and energy, ready to be poured into the molds of stars and galaxies.

That will take time, great stretches of it. The pace of change diminishes, and the cosmic cauldron simmers for thousands of years, losing heat as it continues to grow. Speeding particles, their energy reduced, gather into the larger structures of hydrogen atoms. Immense billows of the gas, blazing a garish yellow, swirl through the reaches of space. As millennia pass, the light suffusing the universe fades to a reddish glow and at last glimmers out completely. Then, about a billion years after the moment of cosmic birth, the darkness is broken. Swarms of stars begin to ignite in the hearts of slowly wheeling hydrogen clouds. These are galaxies in the making, the prototypes of star systems that will ultimately spangle the heavens we see today.

MYTHS OF CREATION

This scenario is the modern successor to the creation myths of antiquity. Some ancient peoples believed that the universe was formed by giants or dragons, or that it began as a primordial acorn or egg. The Greeks spoke of a timeless void that preceded the ordered cosmos: They called it Chaos and told how Gaea, the mother of creation, emerged from that infinite darkness to found the tumultuous dynasty of gods who would rule from Olympus.

Such accounts seem hardly more fantastic than the event that most scientists today believe gave birth to the cosmos and shaped its destiny. Dubbed the Big Bang, their vision of the exploding cosmic seed is almost a comfortable concept in the latter part of the twentieth century, an accepted part of our general kit of knowledge.

But many mysteries remain. Cosmology—the study of the origin and operation of the universe—is a young science, and even its central hypothesis is surrounded by unknowns. For example, no one can say with certainty why the universe popped out of the void. Physicists can only sketch a rough outline of the maelstrom of particles and force that filled the first second of time. Links are missing in the cause-and-effect chain between clumps of particles in the primeval fireball and vast structures of galaxies in the present. And no theory adequately explains just how the diverse forces that determine the physics of our present were once joined together as a single force, as the Big Bang scenario implies.

Searching for solutions to such enigmas, scientists grope through theoretical thickets of appalling complexity, only to find that the cosmos may be the strangest place they could possibly imagine. One hypothesis accounts for the behavior of particles by proposing that the infant universe had eleven dimensions; at a very early point in cosmic history, seven "rolled up," leaving our familiar three dimensions of space and one of time. Another theory laces the present universe with stringlike relics of the early cosmos, weird survivors that are still so stupendously dense that their gravitation attracts vast clusters of galaxies. Others say that the current structure of the universe results from the gravitational effect of nearly massless particles so numerous that a billion are passing through your body as you read this sentence.

That much remains to be learned about the cosmos is hardly surprising. The hypothesis of the Big Bang is only a few decades old. Indeed, the first meaningful strokes in the modern picture of the cosmos were drawn only a few centuries ago, when European astronomers discerned humankind's puny place in the scheme of things.

In 1543, the Polish stargazer Nicolaus Copernicus inaugurated the modern era of astronomy when he proposed a Sun-centered architecture for the universe—a shocking departure from the Earth-centered cosmos that had been accepted since ancient times. His book, *The Revolution of the Heavenly Orbs,* was so original that it reshaped language as well as thought: Henceforth, the word revolution would always carry the connotation of a complete or radical change.

Among the scientists profoundly affected by Copernicus's work were the German astronomer Johannes Kepler and the Italian mathematician Galileo Galilei. Both surveyed the heavens and confirmed the motion of Earth around the Sun. Kepler also demonstrated that all the planets move in elliptical orbits, which can be described in detail with simple mathematical rules that came to be called Kepler's laws. Galileo carried out a set of epochal experiments showing that in the absence of air resistance, all falling bodies, regardless of their size or weight, behave identically. They accelerate—that is, their velocity changes—at a constant, standardized rate.

Kepler and Galileo set the stage for one of the greatest intellectual pioneers of all time, Isaac Newton, born on Christmas Day in 1642, the same year that Galileo died. Newton would absorb scientific learning as though it were the very stuff of life to him, synthesize what had gone before, and then move far beyond his predecessors. For more than two centuries, his theories would be sovereign in defining the operations of the universe.

From the time of his premature birth (at three pounds he was said to be small enough to fit into a quart mug), turmoil figured in Newton's life. His father, a farmer, had died of pneumonia a few months earlier, and his mother struggled to work the family farm at Woolsthorpe, a hamlet about 100 miles north of London. It was a hard time in the countryside. A bloody civil war that would rend England for six years had begun in 1642 at Nottingham, not far from Woolsthorpe. The contending armies of Parliamentary rebel Oliver Cromwell and the Royalist supporters of Charles I regularly advanced and retreated through the small villages.

When Isaac was three years old, his mother remarried, leaving her son to be raised by his grandparents. His early education was in nearby village day schools. At the age of twelve, he was enrolled at the grammar school at Grantham, a town six miles from his home. There he studied Latin—the language of learned discourse in Europe—and the Bible, but he was exposed to little in the way of mathematics or science. Young Newton lived in the house of a William Clarke, the town apothecary, who happened to have one of the best libraries in town and a pretty stepdaughter, with whom Newton later had an adolescent romance, the first and last of his life. He got along poorly with the other boys in the school, who apparently found him strange and altogether too clever.

The quick mind that alienated Newton's classmates found many outlets during the years at Grantham. Years later town residents remembered the mechanical inventions he worked on while other boys played games. Newton built a small windmill from wood. He made a cart that he could propel by turning a crank as he sat on it. He even devised a folding paper lantern that he used to light the way to school on dark mornings. Captivated by the principles of sundials, he learned to tell not just the hour but also the day of the month and to predict such events as solstices and equinoxes. Even the wind fascinated him. One day when Newton was sixteen, a great storm arose. While prudent folks sought shelter from the wind, the young man performed what he later remembered as his debut scientific experiment. He first jumped with the wind, then against it. By comparing the distances of the two jumps, he was able to estimate the force of the gale.

Soon afterward Newton was recalled from school to run the family farm. An old and trusted servant was given the task of teaching him the necessary skills, but Newton never put his heart into the work. Typically, he would build a model waterwheel in the brook—complete with dams and sluices—while his unwatched sheep wandered into the neighbors' cornfields. On market days he bribed the servant to take care of the buying and selling, so that he could

A CHANGING PERSPECTIVE

For millennia, humanity, like a human infant, regarded itself as the apple of the cosmic eye, the sum of the universe and the point around which all else revolved. Although studies of the movements of celestial bodies eventually made some inroads on human self-centeredness, the old notions died hard. Well into the sixteenth century, the prevailing view was that the Sun and other planets orbited the Earth. Then, in the year 1543, Copernicus showed that the Solar System was heliocentric—and the Sun was at once the monarch of stars.

Over the next 400 years, telescopes extended human perceptions into realms that seemed to stretch ever farther into the unknown. First the Milky Way and then distant galaxies were discovered. Gradually, as shown here and on the next two pages, Earth and Sun were removed from center stage to their relatively obscure place: in the wings of one galaxy among billions.

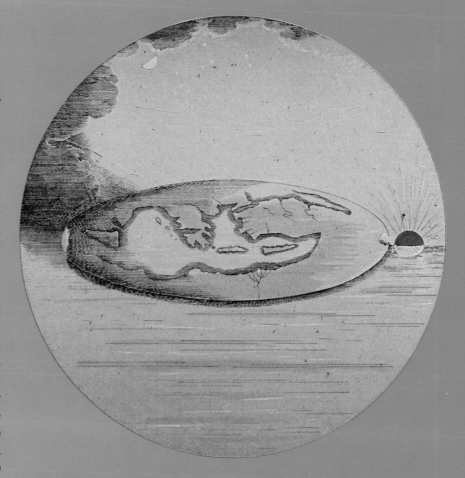

An island Earth. A modern illustration shows the cosmos of the ancient Greek philosopher Thales. His Earth is an island inside a heavenly sphere.

spend his time tinkering with new gadgets or reading. His curiosity, virtually boundless in matters of science and invention, evidently had one limit: It did not extend to farming.

After just nine months, the family decided that perhaps the curious tinkerer would be better off at school. The Grantham schoolmaster, who insisted that the young Newton's talents were wasted on the farm, offered to board him in his own house. So, in the autumn of 1660, Newton returned to Grantham to prepare for the university. By June of the next year, he was ready to go on to Cambridge. Already he aimed at becoming a professor.

Newton paid his way at Cambridge by performing menial tasks for richer students. He may also have turned a profit by loaning out the small allowance he received from his mother. Neither activity earned him many friends. As at Grantham, he was unable to hide his cleverness; moreover, he had adopted an uncommonly puritanical manner at a time when most scholars were discovering the delights of coffeehouses and ale shops. Not satisfied with eschewing these pleasures, Newton even started a coded list of his own

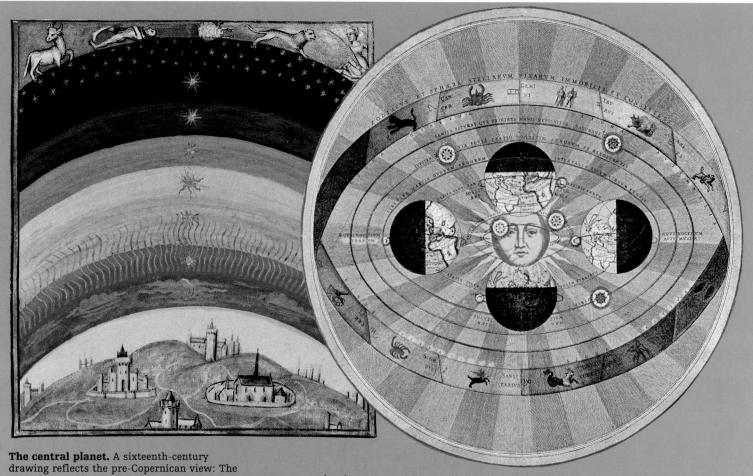

The central planet. A sixteenth-century drawing reflects the pre-Copernican view: The seven known planets travel on separate concentric spheres around Earth.

A dominant Sun. Influenced by Copernicus, the makers of this seventeenth-century print made the Sun, not the Earth, the focal point of an orbiting universe.

sins, which ran to such offenses as "having uncleane thoughts words and actions and dreamese."

At Cambridge, Newton filled his solitude with study on a wide variety of subjects, ranging from astrology to history. By the end of his undergraduate career in 1664, he had also discovered mathematics and natural philosophy—a field that encompassed the subjects now known as the physical sciences.

Newton was preparing to start postgraduate work when his life took another sharp turn. England was struck by the bubonic plague, which took many thousands of lives, mostly in cities like London and Cambridge, whose dirty, crowded confines provided an ideal breeding ground for the rat-borne disease. The university closed temporarily as scholars fled to less-afflicted rural villages. Newton went back to Woolsthorpe, visiting Cambridge from time to time to use the library. At ease in the Lincolnshire heath, he put his prodigious intellect to work on a wide range of problems in science and mathematics, laying the groundwork for a lifetime of accomplishment. He built the first working version of a new astronomical instrument, the reflecting telescope,

Billions of stars. A photographic mosaic shows a region of the Milky Way galaxy, with the galactic center at left. The galaxy's twentieth-century discoverers deduced that the Sun belongs to an assembly of more than 100 billion stars and resides in the outer third of one of the Milky Way's spiral arms.

Hundreds of billions of galaxies. Beyond the Milky Way are billions of other galactic systems whose stars form spirals, ellipticals, and ragged irregulars. Some may be seen in this photograph of an assemblage of galaxies in the Hercules cluster.

which used a curved mirror instead of a lens to focus light. He developed a powerful new branch of mathematics called calculus. And he did the fundamental work on his theory of gravitation.

The popular account of the origin of that theory—that Newton conceived it in the summer of 1666 after watching an apple fall from a tree—is impossible to confirm, but tradition marked a tree on the family farm as the one from which the apple fell. When the tree died in 1820, it was cut into pieces that were carefully preserved. In any event, something during the period turned Newton's thoughts toward the idea of a universal law of gravitation.

BEGINNING WITH MOTION

The foundation for all of Newton's work on gravity was his understanding of motion, which would eventually be expressed as a set of laws. The first of these describes inertia, or the tendency of any body, in the absence of an external force, to stay at rest, or if it is moving, to remain in uniform motion, traveling at a constant speed in a straight line. The second law holds that when a force causes a change in the velocity of an object, the rate of change, or acceleration, is proportional to the force. Galileo's experiments had already shown that falling bodies are in fact accelerating at a constant rate; to Newton, this meant that an object such as an apple was subject to a force of constant magnitude, which he attributed to gravity.

He compared the path of the falling apple to that of an object propelled parallel to the ground, such as a pebble flung from a slingshot. The apple, influenced only by gravity, falls along a straight line that would pass through the center of the Earth. But the pebble, affected by the force of the slingshot as well as by gravity, follows a curved path, moving horizontally even as it falls toward the Earth. Its initial horizontal velocity determines how far it will travel. If a pebble were launched at a sufficiently high speed, Newton reasoned, it would go all the way around the Earth and hit the shooter in the back of the head. And if the shooter ducked, the projectile would continue to orbit the Earth. Newton knew that air resistance would prevent such a thing from actually happening, but he realized that he had hit upon an explanation for why the Moon orbits the Earth. Captured by the gravity of the Earth, it is always falling toward the planet. However, it has enough horizontal velocity to keep it going around forever.

At this point, Newton's thoughts soared beyond the Earth and Moon into interplanetary and interstellar space. The same principles that explain why apples fall to the ground and why the Moon orbits the Earth should also explain why the Earth and all the other planets orbit the Sun. Gravity must be a ubiquitous force that acts between any two bodies in the universe.

Newton arrived at his conclusions by means of a process now known as a thought experiment—using an imaginary scenario to illuminate the rules governing the real world. The procedure is an indispensable tool for scientists in all fields, but especially for cosmologists, whose theories are often impossible to test. Nonetheless, Newton had a powerful means of testing at his disposal: mathematics. He could go far toward confirming his gravity hypothesis by calculating its consequences and then checking his results against observations. First, however, he had to formulate an exact expression of the relations among the gravitational force, mass, and distance.

Newton assumed that the strength of an object's gravitational pull is directly proportional to its mass; that is, the more massive a planet or a star is, the more it attracts another body. Using the mathematical insights expressed in Kepler's laws of planetary motion, Newton deduced that the gravitational force of the Sun on a planet must vary inversely as the square of the distance of the planet from the Sun. Thus, a planet twice as far from the Sun as another of equal mass would experience only one-fourth the attractive force; if an apple were in the same orbit as the Moon—sixty times farther from the center of the Earth than when it is on a tree—it should experience a gravitational pull 3,600 times less. With this inverse-square law, Newton calculated the period of the Moon's orbit, using generally accepted values for the force of gravity at the Earth's surface and for the radius of the Earth. His result, 29.3 days, was far off the mark; the observed period is 27.3 days. Discouraged, he put aside his work on gravitational theory.

Newton's creative flurry culminated with his return to Cambridge in 1667, at the age of twenty-five. Bringing with him his new telescope and some new mathematical theorems, he settled back into the scholarly routine of the university, where his genius at last began to win him more than the scorn of less-gifted classmates. Isaac Barrow, who occupied the prestigious Lucasian chair of mathematics, was so impressed that when he resigned in 1669, he recommended Newton as his successor. The new professor quickly won renown among his scientific colleagues both for his creative experiments on the properties of light and for his ability to solve difficult mathematical problems with extraordinary speed.

THE PRICE OF FAME
An unexpected side effect of his spreading fame was a blizzard of scholarly challenges. Admitted in 1672 to the Royal Society of London, where distinguished scientists met to discuss the latest discoveries, Newton suddenly found himself embroiled in seemingly endless quarrels that drove him to distraction. A particular tormentor was Robert Hooke, an ingenious and argumentative natural philosopher who was responsible for arranging weekly scientific demonstrations for members of the Royal Society. These were so draining that Newton soon despaired of learned discourse, "for I see a man must either resolve to put out nothing new, or to become a slave to defend it."

A wary Newton thus avoided going public even when new observations seemed to offer confirmation of his languishing theory of gravitation. In the same year that he joined the Royal Society, the group was told that a French astronomer named Jean Picard had remeasured the diameter of the Earth and found it to be about 15 percent larger than previously thought. When Newton incorporated the revised measurements into his gravitational equations, the

THE RIDDLE OF
THE NIGHT SKY

For centuries, philosophers and scientists puzzled over a deceptively simple question: If the universe is infinite and full of stars, how can the sky be dark at night? Just as an observer in a forest sees trees in all directions *(opposite)*, every line of sight for an observer in an infinite universe should end with a star. The net effect should be a continuous background of celestial light—what literary genius Edgar Allan Poe called the "golden walls of the universe." Not only would the night sky be far brighter than ordinary day but also the heat radiated by all those stars would evaporate the oceans and the Earth itself.

Since this is clearly not the case, early theorists speculated that either stars were limited in number after all or their light somehow failed to reach Earth.

When astronomers discovered interstellar dust, some thought the culprit had been found. But calculations quickly showed that if dust particles absorbed the energy of all the missing starlight, the dust itself would begin to glow.

The eventual answer came out of two widely accepted implications of relativity theory. First, if the universe began in a so-called Big Bang a finite amount of time ago, as its observed expansion suggests, and second, if light travels at a finite speed, only the light of stars within a given distance has had enough time to reach the Earth *(below)*; any stars farther away are simply undetectable. Thus, even if the number of stars is infinite, the number of visible stars is not, allowing for dark gaps in the sky. Other effects also add to the darkness. Over time, for example, stars burn out, creating new gaps. And in the expansion of space following the Big Bang, light itself undergoes a transformation in transit *(overleaf)*.

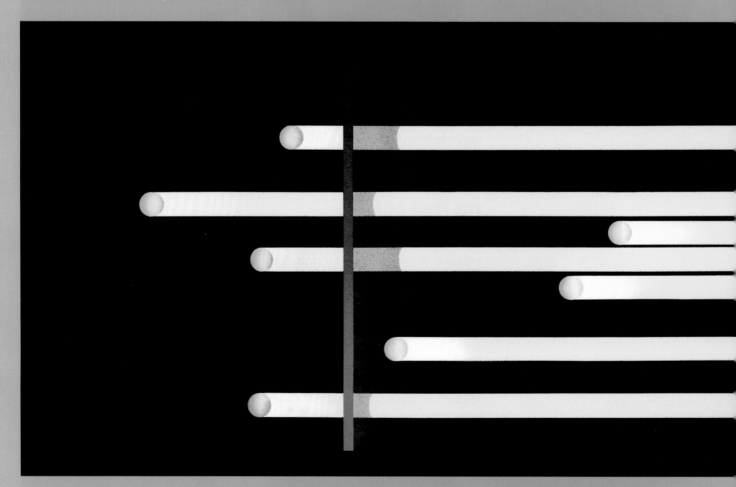

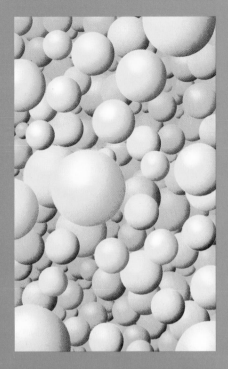

An observer in the midst of a forest *(far left)* would perceive a screen of trees because the forest continues farther than its so-called background limit—the average distance at which the observer's line of sight intercepts a tree. Similarly, from any point in a boundless, star-filled universe, near and distant stars should overlap every inch of the field of view *(left)*, filling the night sky with light.

One reason the night is dark is that the universe began a finite period of time ago. If the cosmos is 15 billion years old, for example, an observer on Earth can detect only light from objects within 15 billion light-years, a distance indicated below by a translucent plane. Stars on the far side of that mark are invisible, for their light simply has not had time to travel far enough.

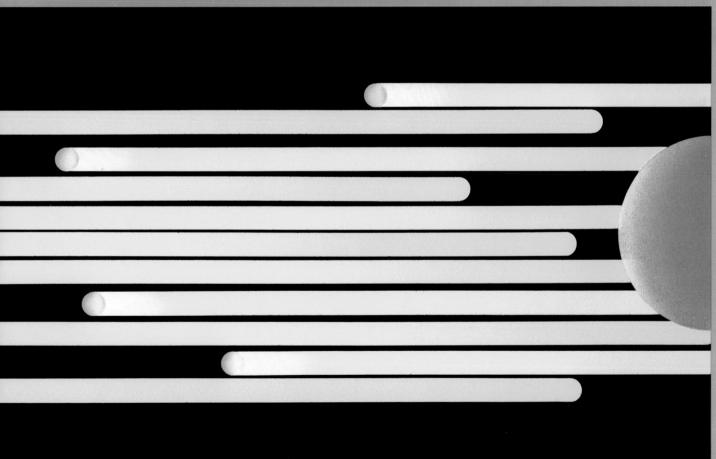

Light radiating from a stationary source looks the same to observers in all directions, provided they are also stationary.

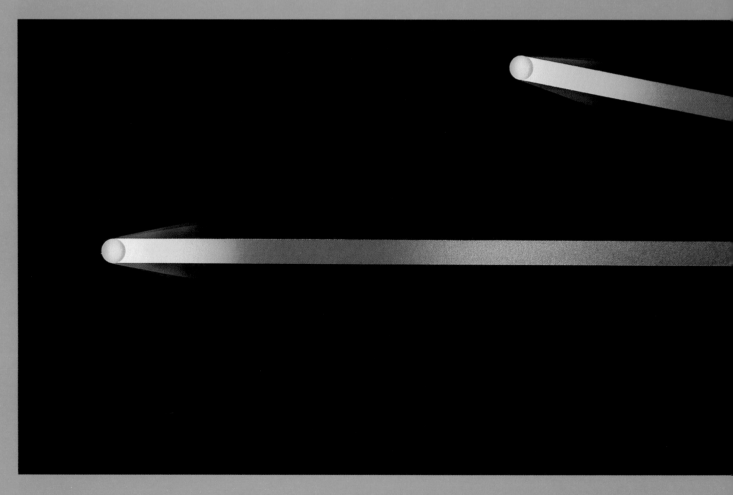

Shifting from Light to Darkness

While the number of stars visible from Earth is curtailed by an observation limit of several billion light-years *(pages 20-21)*, starlight is also effectively reduced by "cosmological red-shifting," a phenomenon that is directly related to the universal expansion implied by relativity theory.

Cosmological red shifts are similar to the classic Doppler effect, for both can involve the stretching of light waves. In the Doppler version, illustrated at the top of the next page, the motion of a light source relative to an observer produces the change. Light waves from an object moving toward the observer become compressed toward the higher frequency, or blue, end of the spectrum; waves from an object moving away are stretched and shifted toward the lower frequency, or red, end.

In the cosmological case, red shifts have nothing to do with the motion of a light source, but rather with the source's distance from the observer. According to relativity's equations, space itself is expanding everywhere and in all directions, except within gravita-

The Doppler effect shifts light from a receding source toward red and light from an approaching source toward blue.

tionally bound systems such as galaxies. Light from a faraway source in a distant galaxy thus travels a stretching path that elongates each wavelength. Just as in a Doppler red shift, the stretched wave results in a lower-frequency, redder form of light. The more remote a galaxy is from Earth, then, the more intervening space there is to expand, the more rapidly the distance to that galaxy grows, and the greater the red shift is. The light of stars in relatively nearby galaxies will simply redden, but over greater and greater distances starlight may be shifted past the low-frequency end of the visible spectrum into the invisible zones of infrared and then radio.

The expansion of the universe dims starlight from very distant galaxies through a cosmological red shift—analogous to the Doppler effect *(above)*—that alters the light emanating from receding objects. As shown below, the more distant a galaxy, the longer the path its light must travel to reach Earth. Moreover, because the distance between the galaxy and Earth is constantly increasing, each wavelength of starlight is stretched and its frequency lowered. Light from the stars in far-off galaxies can thus shift completely out of the visible realm, reducing to some extent the starlight that is received on Earth.

calculated orbital period of the Moon—once two whole days off the mark—came into agreement with observations.

But Newton remained reticent. Instead of announcing his breakthrough, he continued his solitary ways, living a cloistered life at Cambridge. His lectures as Lucasian Professor of Mathematics were so sparsely attended and poorly understood that, Newton later lamented, he often "for want of hearers, read to the walls." During this period, he was also caught up in new enthusiasms, producing several papers on such subjects as alchemy and theology. He even developed plans for a multivolume history of the early Christian church.

Nevertheless, the thread of his abiding passion was unbroken. He maintained a correspondence with other scientists, including his old nemesis Hooke. This circle of colleagues continued to be fascinated by the problems of gravitation and planetary motion. In one discussion in 1684, involving Hooke, the astronomer Edmund Halley, and the celebrated architect Christopher Wren, Hooke claimed to have produced a mathematical proof that Kepler's laws could be explained by an inverse-square law of attraction. Hooke was known for making extravagant claims about his work, and he backed down when Halley and Wren pressed him for details. Soon afterward, during a visit to Cambridge, Halley asked Newton what path a planet would take around the Sun if the Sun's attraction decreased by the square of the distance.

"An ellipse," Newton promptly replied. When the amazed Halley asked how he knew, Newton's response was immediate. "Why, I have calculated it." Halley asked to see the calculations, but Newton, never a paragon of tidiness, could not find his notes from nearly two decades before. Instead, he had to redo the work and send it to Halley, who urged Newton to publish these and other computations in a book. Only when Halley offered to pay for the publication did the ever-reluctant Newton agree.

Thus began two years of intense concentration on some of the most daunting problems in science. To support the broadly drawn concepts of his gravitational theory, Newton had to show how they operated in specific situations, which required that he work out many details. For example, he had to have a standard way of defining the distance between two celestial bodies so he could figure their gravitational attraction; should the measurement be taken between the surfaces of the bodies or between their centers? It was an arduous mathematical labor to show that the attraction between two large spheres such as the Sun and the Earth could be accurately represented by pretending that all the mass of each sphere is concentrated at its center.

So unrelenting was his focus on these calculations that Newton became the epitome of the absent-minded professor. He dressed carelessly, forgetting to wear socks, tie his shoes, or comb his hair. At times he even forgot to eat or sleep. When at last the book was published in 1687, he was exhausted and ill. But his completed work, *The Mathematical Principles of Natural Philosophy*—or the *Principia,* to use the shortened version of the Latin title—was, beyond any doubt, a titanic achievement. In addition to announcing a universal law of gravitation and laws of motion, the *Principia* contained page after page of astonishing mathematical analysis that not only explained the orbits of the planets and comets around the Sun and the orbit of the Moon around the Earth but also showed how ocean tides are caused by gravitational effects. Newton had computed the mass of the Sun, the Earth, and Jupiter. He had explained the observed flattening of the poles of Jupiter as a result of rotation and had predicted flattening at the Earth's poles. He had even included a discussion of a method for launching artificial satellites.

The first edition sold out so quickly that some scientists, reluctant to wait for a second printing, reportedly copied the entire book by hand. Overnight, Newton became a public hero, "the great ornament of the present age," in the words of a contemporary editor. Another writer praised Newton as an "Honour to his Country, and an Advancer of the noblest Learning, and an Enlarger of the Empire of the Mind."

The acclaim finally drew him away from the cloistered life in Cambridge. He continued his scientific work, but in 1696, when he was appointed warden of the mint, with responsibility for the coining of new currency, he came into a substantial income and moved to London to enjoy his celebrity. In 1703 he was elected president of the Royal Society, and two years later he was knighted by Queen Anne.

HOLES IN THE THEORY

As effective as the gravitational theory was in explaining the dynamics of the universe, its author recognized its shortcomings. "To explain all nature is too difficult a task for any one man or even for any one age," Newton acknowledged. He had particular trouble trying to understand the actual nature of gravity and space. While his theory predicted the effects of gravity quite handily, it said nothing about the mechanism by which gravity acts. In fact, Newton believed that gravity resulted from divine action; in effect, a stone fell because God's finger was pushing it.

In his mathematical explanations for the planetary motions, Newton had worked on the assumption that gravity operates instantaneously throughout space. He disliked this idea, but could think of no alternative; in any event, it seemed to have no practical importance for his calculations. The speed of gravitational action did, however, turn out to be of critical importance when the universe was considered as a whole. Newton believed that the universe was infinite. Otherwise, he argued, it would have an edge and therefore a center of gravity, like any other finite object. The attraction between its parts would, as he put it, cause the universe to "fall down into the middle of the whole space"—which clearly has not happened. By contrast, each piece of matter in an infinite universe is subject to equal forces from every direction and therefore stays put.

Newton was a little bothered by the fragility of a universe governed by the balance of these opposing forces. If gravity acts instantaneously over infinite distances, then the forces on each piece of matter would be not only equal but also infinite, in every direction. Any tiny imbalance in the distribution of matter would upset the equilibrium of attraction, subjecting bodies to enormous asymmetrical forces, far stronger than the ordinary gravitation keeping planets in their orbits or holding stars together. The consequences would be catastrophic: Planets would be yanked from their paths and pulled through interstellar space at incredible speeds. Since the universe seemed to be sticking together, however, Newton concluded that the distribution of matter was in fact perfectly uniform and that the net gravitational effect of distant bodies was zero.

Another subject arose to puzzle Newton as he considered the circumstances of bodies subjected to no external forces. The law of inertia, first stated by Galileo and appropriated by Newton as the first of his laws of motion, states that an object continues in a state of rest or of uniform motion unless it is compelled to change its motion by forces acting on it. But there is no clear standard by which to judge whether an object is at rest. For example, a passenger on a ship on a perfectly calm night might see lights moving past in the darkness. Those lights could be interpreted as a sign that the ship is moving forward past another, stationary ship, but they could equally mean that the observer's ship is at rest while another ship is passing. Or both ships could be moving. As long as the motions involved are uniform, the stationary condition cannot be determined.

Newton solved this abstract problem with another abstraction. An object is at rest, he declared, if it has no motion when compared to "absolute space," which "remains always similar and immovable." He pictured absolute space as an invisible grid, against which any motion could be plotted. He was not disturbed by his conclusion that human observers could never distinguish between absolute motion and rest; it was sufficient to him, as a devout Christian, that God be able to tell the difference.

In any event, Newton's theory worked very well in accounting for the behavior of things humans could perceive, from cannonballs to comets. One of the most sensational successes of Newtonian physics came in the mid-nineteenth century. Observed irregularities in the orbit of the planet Uranus led two young mathematicians, Urbain Leverrier in France and John Adams in England, to a startling conclusion: There must exist another, fairly large planet more distant than Uranus. Working independently, they used Newton's laws of motion and gravitation to compute the new planet's position. In September of 1846, Neptune was discovered at just the place where Leverrier and Adams had forecast. Newton's law of gravitation seemed a scientific tool of matchless power and perfection.

Attempting to prove the existence of the ether, an invisible medium filling space, American scientists Albert Michelson and Edward Morley conducted an ingenious experiment in 1887. As shown in the simplified diagram above, a light source (A) sent a beam of light to a thinly silvered plate (B) that allowed some of the light to pass through and reflected the rest at a right angle. The two beams then traveled equal distances to mirrors (C and D) that reflected them back to the plate. The light joined again into a single beam to enter the telescope (E). There, a device measured light-wave patterns to determine whether the two beams arrived simultaneously.

The scientists believed that if the Earth was moving through the ether, the light beam traveling upwind would be slightly slower than the one traveling crosswind. However, the telescope indicated that both traveled at the same speed. The unexpected result later helped Einstein to realize that the speed of light is a constant.

OF ELECTRICITY AND MAGNETISM

By the latter half of the nineteenth century, physicists—believing that gravity and motion were settled subjects—began turning their attention to two other, more mysterious phenomena, electricity and magnetism. These were known to be related: A moving electric charge could produce a magnetic force that would deflect a nearby compass needle, and a moving magnet could produce an electric current in a neighboring wire. But while electricity and magnetism were clearly two sides of the same coin, no one knew just what the coin was.

In 1865 James Clerk Maxwell, a Scottish physicist, published a mathematical description of the relation between electricity and magnetism. His formulas, now called Maxwell's equations, showed that a vibrating object with an electrical charge will radiate electromagnetic waves, analogous in many ways to the waves that spread out from a pebble tossed into a pond. The equations predicted that the speed of these waves should be 186,000 miles per second—exactly the speed of light as already determined by various experiments. Maxwell concluded that these electromagnetic waves were similar to light, which was known to have a wavelike nature. In fact, he decided, visible light was simply one of many forms of electromagnetic energy, distinguished from the others only by its different wavelength.

The theory of electromagnetic waves exacerbated a dilemma that had first arisen when experiments indicated the wave nature of light. Physicists of the day believed that all waves required some medium to carry them, as water carries the ocean swell. But space, through which the light of stars travels, was generally considered to be a vacuum. The favored solution was to postulate the existence of a wave-carrying medium called the ether, an insubstantial, invisible stuff that did not impede the motion of celestial bodies.

In 1887 two American scientists, Albert Michelson and Edward Morley, conducted an experiment to detect the ether. Their instrument, developed by Michelson, used the principle of interference of light waves—the strengthening or weakening of waves that are out of phase—to measure the speed of light in different directions. If the Earth is moving through the ether, they reasoned, then a light beam pointed along the direction of the Earth's motion should travel at a different speed from that of a beam moving perpendicular to the motion: The Earth's motion and the motion of the forward-pointed light should add together, whereas the other beam would gain no boost from the Earth's motion. Michelson and Morley were stunned when their experiment indicated that regardless of which direction the light beam was aimed, its speed was the same *(opposite)*. Convinced that their equipment must be at fault, they repeated the experiment with even greater precision. But the results were identical, forcing them to a conclusion completely at odds with common sense: The speed of light is not influenced by the motion of its source or the motion of an observer. It is always the same.

For the next two decades physicists struggled in vain to reconcile these findings about electromagnetic waves with the definition of space inherent in Newton's laws. If space and time are absolutes, then it is not possible for two observers, one moving and one at rest, to perceive the same light beam as moving at the same velocity relative to themselves. But that was precisely what the Michelson-Morley results implied. Apparently either Newton or Maxwell was wrong, even though each of their theories seemed to work perfectly in describing everything else it was applied to.

YOUNG EINSTEIN

The way out of this quandary began to take shape in 1895 in the mind of a sixteen-year-old schoolboy. Albert Einstein, born in Ulm, Germany, found his life in upheaval when his family moved to Italy in 1894, after the failure of his father's Munich electrical firm. Left in Munich to finish the school year, Albert soon decided to drop out and join his family. In Italy he spent a year hiking, thinking, and studying on his own. During this period he began to contemplate the effects of moving at the speed of light—a puzzle whose solution would change physics and cosmology forever.

Einstein's insatiable curiosity had appeared at an early age. In his autobiography he recalled his awed pleasure over a magnetic compass his father brought him when he was ill in bed as a small child: a needle, isolated and untouchable, yet caught in the grip of an invisible force. He never lost this childlike sense of wonder, in spite of what he described as the "dull, mechanized method of teaching" in the Munich schools, where discipline was instilled by such means as hitting students on the palms of their hands with a ruler during counting and multiplication exercises.

Young Einstein's attitude toward the schools was reciprocated by instructors like his Greek teacher, who told him that he would "never amount to anything." Since the German schools stressed Greek and Latin studies instead of science and mathematics, Einstein's poor memory was more often in evidence than his flair for problem solving, and many considered him dull-witted. After his year in Italy, the youth moved to Zurich, Switzerland, in hopes of attending the prestigious Federal Institute of Technology, known as the Poly. He failed the entrance exam and spent a year in a Swiss high school before successfully reapplying. While Einstein soon found that most of the courses bored him, other aspects of life in Zurich were more congenial; he made many friends and began to think of Switzerland as his home.

One of his new friends was classmate Marcel Grossman, a bright, hardworking mathematics student whose diligent habits were the opposite of Einstein's. When Einstein was in danger of flunking his graduation examinations because he had cut so many classes, Grossman, who had attended all the classes and taken careful notes, loaned the notes to his friend. Einstein crammed for the exams, passed them, and graduated in 1900.

With graduation came the end of an allowance from his family, and Einstein had to look for a job. Without recommendations—he later recalled that he was "not in the good graces of any of my former teachers"—he could not find permanent work and had to make ends meet with tutoring and part-time teaching. After two years of sporadic employment, Einstein

Albert Einstein, shown here at his desk at the Swiss Patent Office in Bern, published his first paper on relativity while he was working as a patent examiner in 1905. The young physicist spent seven years examining patent applications for technical errors.

again benefited from his friendship with Marcel Grossman, who by this time was teaching mathematics. Through a family connection, Grossman was able to secure Einstein a position as a technical expert, third class, at the Swiss Patent Office in Bern.

The next few years were full and productive. He married Mileva Maric, a classmate at the Poly, in 1903, and their first child, Hans Albert, was born in 1904. He quickly mastered his job of examining patent applications, and bootlegged time for his own studies of such subjects as the physical properties of light. In the evening he worked on science or invited friends to his apartment to discuss physics, philosophy, and literature. Typically these meetings were lively and loud, lasting far into the night, much to the annoyance of the neighbors. Though Einstein was essentially a loner, the opportunity to develop ideas and try them out on the keen intellects of his friends was invaluable. He began to publish the results of his research in one of the leading scientific journals, and he focused his intuitive and analytical powers on the implications of the question that had intrigued him years earlier: What would it be like to ride a light beam?

Einstein first considered the situation of an observer at rest, with light waves passing by. The observer would see a regular pattern of crests and troughs moving through space. If the same observer were to accelerate to match the speed of that beam, the wave pattern would presumably disappear. The Maxwell equations, however, require that electromagnetic waves maintain their wave nature, no matter what the action of an observer. Thus, either the equations must be wrong or it must be impossible for an observer to move at the speed of light. But Maxwell's theory worked fine in every real-life application, and classical physics contained no prohibition against moving at the speed of light—or even faster, for that matter.

Einstein proposed a revolutionary way out of this dilemma in a 1905 paper in the prestigious German journal *Annals of Physics.* The epochal article, called "On the Electrodynamics of Moving Bodies," introduced the principle that became famous as relativity. Einstein began by postulating the speed of light as a constant, regardless of the motion of the source or the observer. To eliminate the resulting conflict with classical physics, Einstein extended Newton's ideas about the physics of motion. Newton had, in effect, invalidated the concept of absolute uniform motion, showing that the only detectable state of motion is one where an object is moving relative to an observer. Einstein's leap was to invalidate absolute space and time as well. According to his theory, the dimensions of an object and the duration of an event are not fixed values. Rather, they may be determined only by considering the motion of their frame of reference relative to an observer. As long as that motion is only a small fraction of the speed of light, the changes in space and time would be nearly imperceptible. In a vehicle moving at nearly the speed of light, however, the changes would become very apparent. A light beam would still travel at its constant speed, but by the standards of an observer at rest, the yardstick used to measure the distance the light traveled would be shorter and the clock used to time the light's passage would run more slowly *(pages 39-57)*.

Building upon these foundations—the constant speed of light and the relativity of space and time—Einstein used comparatively simple algebra to extend the theory. He was able to show that when an object moves at speeds close to that of light, its mass increases in proportion to its kinetic energy. From this relationship between mass and energy, he deduced that the two properties are interchangeable—a conclusion that he expressed in the famous equation $E = mc^2$, where E is the energy content of an object, m is its mass, and c is the speed of light.

The equation was central to the first test of relativity, based on experiments begun in 1901, before Einstein proposed the theory. German physicist Walter Kaufmann, measuring the mass of high-energy radioactive particles, had detected no change in mass when the particles were accelerated. Einstein, however, was unperturbed when he learned of Kaufmann's results. He was convinced that the underlying assumptions of the theory were valid and that the theory itself was so sensible that it had to be right. This faith in his intuition was typical of him, and it was well justified. When the radioactive-particle experiments were repeated by other scientists over the next ten years, the initial results were found to be in error. The theory was upheld.

Renowned scientists such as the German physicist Max Planck praised Einstein's work; Marie Curie, winner of a Nobel prize in 1903 for the discovery of radium, called the young physicist "one of the leading theoreticians of the future." Einstein was finally able to leave the Patent Office in 1909 for a position at the University of Zurich (but this occurred only after the first choice for the post withdrew, arguing that Einstein's scientific promise was far greater than his own). Eager as he was to gain more time for research, Einstein left the Patent Office with some regret. In a letter to a friend, he called Bern "that secular cloister where I hatched my most beautiful ideas and where we had such good times together." Over the next few years, he moved frequently to ever more attractive university positions. Finally, in 1914, through the efforts of Max Planck, he reached the top of his profession, accepting positions as a research professor at the University of Berlin and as director of the Kaiser Wilhelm Institute for Physics, a major research center still under construction in Berlin.

By this time Einstein had been working for nearly a decade to broaden the scope of his theory. As originally stated, it applied only to systems in uniform motion—moving relative to one another along straight lines at constant speeds. Einstein wanted to generalize the theory to include nonuniform mo-

The Forces of Nature: A Cosmological View

Central to the realm of cosmological theory are the four known forces of nature—gravity, electromagnetism, and the subatomic strong and weak forces. Collectively referred to as field forces, the quartet differs fundamentally from the familiar mechanical forces of everyday experience.

The ordinary notion of force involves a tangible agent acting directly on some object, as in the case of a horse harnessed to a wagon. The horse pulls; the wagon moves. Scientists explain gravity and the other field forces in another way. The fall of an apple is not the result of a mechanical force transmitted by the Earth through some invisible harness. Instead, the apple moves because of its interaction with a local gravitational field that is created by the mass of the Earth.

The field *is* gravity; at every point in space, it has a magnitude that can be stated in terms of the force it exerts on an object placed there. Earth's gravitational field, for example, is weaker atop a mountain than at the bottom of the ocean.

The movement of an object through a field creates a complex situation. For instance, when a charged particle traverses an electromagnetic field, it induces changes in the field. The altered field, in turn, subjects the particle (and all others under its influence) to continuously varying force levels.

Scientists sort out such intricate dynamics by recourse to mathematical expressions called field equations, the underpinnings of force theories. Because these equations also make it possible to calculate previous characteristics of a field, they are invaluable to cosmologists. By backtracking through the interactions of matter and force fields, theoreticians can draw ever more accurate pictures of the universe as it was in its infancy.

tions, amplifying the laws of physics so they would be the same for all systems, accelerated as well as uniform. To do this, he had to examine the relation between the acceleration produced by gravity and that generated by other forces. He started with a thought experiment.

Einstein imagined a "spacious chest resembling a room" drifting through space at a constant speed, far from any stars or planets. The contents of the chest, including an observer, would be weightless, floating freely; there would be no up or down, no floor or ceiling. If a constant force were applied to one side of the chest, it would accelerate uniformly, and one wall would begin to push against the contents. To the observer inside the chest, that wall would become the floor. The accelerating force could be adjusted so the observer would sense a downward pull identical to that felt by someone on the surface of the Earth. The observer in the chest would in fact have no way of knowing whether the pull was the product of gravitation or acceleration. Therefore, said Einstein, gravity and acceleration are equivalent.

This principle of equivalence first appeared in a paper Einstein published in 1911 and was the fundamental premise of the theory of general relativity. (Einstein's earlier theory would become known as special relativity.) The full version of general relativity would appear four years later. Einstein used the intervening time to develop mathematical descriptions of the interaction between matter, radiation, and gravitational forces. In contrast to the calculations he had used to support special relativity, these field equations of general relativity—so called because they described the nature of gravitational fields—were complex and stretched his mathematical capacity to the breaking point. In desperation he wrote to his old friend Marcel Grossman, "Help me, Marcel, or I'll go crazy!"

Grossman, by now a professor of mathematics at the Poly in Zurich, was an expert in non-Euclidean geometry, the geometry of curved surfaces and spaces. He introduced Einstein to the work of the nineteenth-century German mathematician Bernhard Riemann, who, with no particular application in mind, had developed the computational tools that Einstein needed. Grossman helped him mine these waiting treasures.

Einstein required non-Euclidean geometry because general relativity was all about curves. The theory predicted that any massive object would, by its gravitational attraction, bend the path of anything passing near it. This was no surprise in the case of matter, of course; the curved trajectories of planets and comets were well known and seemed to be fully explained by Newton. But Einstein maintained that gravity would also bend the path of electromagnetic waves. A light beam, for example, would curve as it traversed the neighborhood of a massive star.

General relativity was meant to provide a single explanation for all physical phenomena, including the way light follows the shortest path between two points. In Euclidean geometry, this path is a straight line, not a curved one. Working out his theory with non-Euclidean geometry, Einstein described an entirely new way of looking at the universe. Instead of saying that light rays curve, he said that the fabric of space itself is warped by gravity.

The concept of curved space, based as it is in complex mathematics, may best be illustrated with a two-dimensional analogy: the surface of a sphere. The geometry of flat surfaces does not apply to curved surfaces. For example, parallel lines drawn on a flat surface never meet, but they always intersect when drawn on the surface of a sphere. Indeed, the concept of a straight line has no meaning on this surface, because all lines have a certain amount of curvature. As all navigators know, the shortest distance between two points on the Earth's surface is a segment of a great circle, a line traced on the surface by a geometric plane passing through the center of the Earth. A more general term that refers to the shortest path between two points on any curved surface is a geodesic, from Greek words meaning "divide the Earth."

General relativity describes the universe in terms of four dimensions, three in space and one in time, with gravity a geometrical property of this four-dimensional space-time. Newton's view had been that the mass of the Sun suffuses the space around it with a gravitational force that makes the planets move along curved trajectories instead of straight lines. Einstein described the same phenomenon by saying that the mass of the Sun warps space-time; planets follow a certain orbital track not because the Sun attracts them but because they are moving along geodesics, the shortest paths in curved space-time.

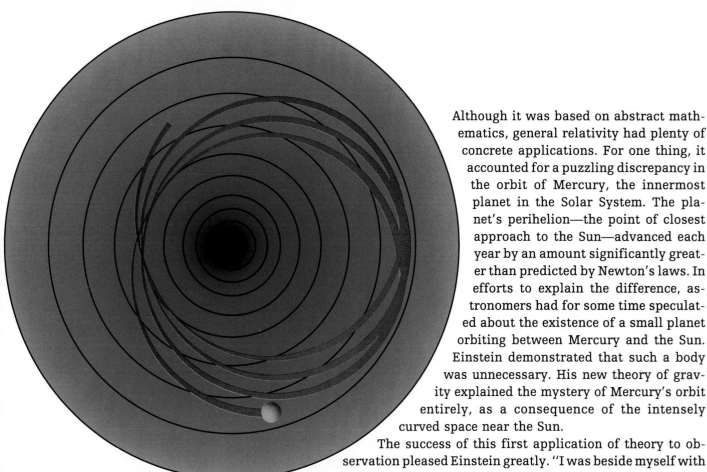

According to Newtonian theory, Mercury's perihelion—the point at which its orbit is closest to the Sun—will advance slightly every year because of the gravitational influence of other planets *(above)*. This shift, or precession, should amount to 531 seconds of arc per century. (An arc second, 1/3,600 of a degree, is an angular measure of the sky.) But Mercury advances 574 seconds of arc per century—leading baffled scientists to speculate on the existence of another planet closer to the Sun.

Einstein's general theory of relativity explains the 43-second discrepancy: Massive objects, such as the Sun, literally warp the space around them, creating a depression, or gravity well, in space. As the planets move around the Sun, their orbits follow the outlines of the well, and it takes them a tiny bit longer to follow that sinking path than it would if they followed the flat paths of a Newtonian system. Mercury, the closest planet to the Sun, shows the difference most clearly.

Although it was based on abstract mathematics, general relativity had plenty of concrete applications. For one thing, it accounted for a puzzling discrepancy in the orbit of Mercury, the innermost planet in the Solar System. The planet's perihelion—the point of closest approach to the Sun—advanced each year by an amount significantly greater than predicted by Newton's laws. In efforts to explain the difference, astronomers had for some time speculated about the existence of a small planet orbiting between Mercury and the Sun. Einstein demonstrated that such a body was unnecessary. His new theory of gravity explained the mystery of Mercury's orbit entirely, as a consequence of the intensely curved space near the Sun.

The success of this first application of theory to observation pleased Einstein greatly. "I was beside myself with ecstasy for days," he wrote a friend. The feat also impressed his scientific colleagues, but it was after all an explanation of facts that were already known. A far more convincing proof of theory would be the confirmation of the predicted bending of light—specifically the bending of starlight by the Sun. This effect could only be observed during a solar eclipse, when the relative positions of a group of stars near the Sun could be photographically recorded. These positions would then be compared with the relative positions of the same group six months earlier or later, when the Earth was between the Sun and the stars.

Einstein was lucky in the timing of efforts to detect this effect. His first calculations, published in 1911 before he had fully developed his theory, predicted a deflection of 0.83 second of arc, equivalent to the width of a pinhead when viewed from a distance of 1,000 feet. This was only about half the value that he arrived at in the finished theory. In the interim, however, two expeditions were mounted to check his wrong prediction. Happily for Einstein's reputation, neither mission achieved an observation. An Argentinian eclipse expedition to Brazil in 1912 was rained out, and a German effort in 1914 was canceled at the outbreak of World War I. In 1916, working with the theory's full mathematical framework, Einstein predicted a deflection of starlight equal to 1.7 seconds of arc. General relativity was ready for a make-or-break test.

Meanwhile, even though the war disrupted communications in Europe, news of Einstein's work had spread quickly. The physicist—who had become a Swiss citizen in 1901—had friends and admirers in the scientific community of the Netherlands. Among them was astronomer Willem de Sitter, who got his hands on the completed theory and immediately set to work applying it to astronomical problems. In late 1916 he submitted the first of three papers to the Royal Astronomical Society of England for publication in their journal, along with a copy of one of Einstein's papers on general relativity. The secretary of the Royal Astronomical Society, responsible for finding reviewers for the papers, was Arthur Eddington of Cambridge University, a brilliant astrophysicist. German research was the object of disdain and outright hostility among British scientists, but Eddington, a Quaker and a pacifist, did not share this prejudice. He took it upon himself to give the papers a fair reading.

In view of the theory's difficult mathematics, this was not a task lightly begun. When Eddington finished, he wrote that general relativity "claims attention as being one of the most beautiful examples of the power of general mathematical reasoning." In an exceptional display of moral and intellectual courage, he set aside his own research to explain, promote, and defend the insights of a scientist working in the capital of Britain's mortal enemy.

Eddington soon found an ally in Frank Dyson, the British Astronomer Royal and someone with influence in high places. The two began to make plans for an eclipse expedition to test the theory. Dyson had more than a purely scientific interest in the project: He was trying to keep Eddington, one of Britain's finest minds, out of a labor camp. As an able-bodied single man of thirty-four, Eddington was eligible for the draft, but he had declared his

Relaxing in his garden, English astrophysicist Arthur Eddington *(near left)* shares a bench with his colleague Albert Einstein *(far left)* in 1930. Twelve years earlier, Eddington had written the first complete account in English of Einstein's general theory.

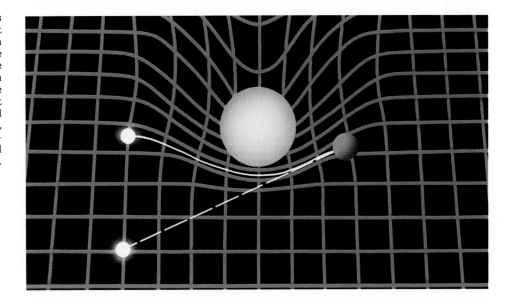

A simplified model illustrates Einstein's statement that light follows a curved path near a massive object. Space can be viewed as a sheet stretched by the heavy Sun *(yellow)*. As light from a distant star *(solid line)* passes the Sun on the way to Earth *(blue)*, it will follow the sheet's warped surface. Viewed from the planet, the star's apparent position *(dotted line)* will differ from its real position by a predictable amount.

intent to seek deferment as a conscientious objector. He was quite prepared to accept the consequences—confinement to a work camp, social disgrace, and possibly professional ruin, as well as acute embarrassment to Cambridge University. But Dyson had other ideas. He arranged that when Eddington was deferred, he would not be sent to a camp; instead, he would undertake an arduous trip to observe a full solar eclipse.

The eclipse that was chosen would occur on May 29, 1919, when the Sun would be passing in front of the Hyades, a rich cluster of stars in the constellation Taurus. Two groups would perform the observations: One, led by Eddington, would trek to a cocoa plantation on the small island of Príncipe off the west coast of Africa; the second, led by two other British astronomers, would go to the village of Sobral in Brazil.

The necessary instruments were impossible to obtain during the war, but as soon as an armistice was signed in November of 1918, preparations were made in great haste. The following March, the two groups set sail. In Príncipe, eclipse day brought rain and a cloud-covered sky. Eddington despaired. Then, as the eclipse neared totality, there was a break in the overcast, and the team was able to get two photographs showing the desired star images.

Eager to know the answer, and as a precaution against a mishap on the way home, Eddington immediately began to analyze the photographs. "Three days after the eclipse," Eddington wrote, "as the last lines of the calculations were reached, I knew that Einstein's theory had stood the test and the new outlook of scientific thought must prevail." The group in Brazil was more patient, bringing its photographs back to England for examination. In the months that followed, the data from both expeditions were carefully evaluated. Eddington's preliminary conclusion was confirmed.

Despite the armistice, communication between England and Germany was still difficult, so Einstein learned of his success indirectly, from his friend and colleague Hendrik Lorentz in the Netherlands. He then sent a postcard to his gravely ill mother in Switzerland: "Dear Mother, Good news today. H. A.

Lorentz has wired me that the British expeditions have actually proved the light deflection near the Sun."

On November 6, 1919, Dyson made it official in a public report to the Royal Society of London. In a world weary of war, science offered humankind a new and amazing way of looking at the universe. Albert Einstein became, in the words of the *New York Times,* "the suddenly famous Doctor Einstein."

NO EDGE TO THE COSMOS

Very little affected by his new celebrity, Einstein had already begun to explore the implications of general relativity for the universe as a whole. Like Newton, he rejected out of hand the idea of a universe with a boundary. His reasoning was based on the tenet that "all places in the universe are alike," a widely shared assumption that came to be known as the cosmological principle. The assumption is that the universe looks generally the same from any vantage point. No part is exactly identical to any other part, but the overall makeup of the universe is the same everywhere. Observers on Earth should see as many stars in the skies of the Northern Hemisphere as in the heavens of the Southern Hemisphere, and observers far away in space should see a similar celestial picture. By contrast, the view near the edge of a bounded universe would be radically different from what would be seen near the center.

If the universe has no boundary, the only apparent alternative is an infinite universe. But Einstein, using non-Euclidean geometry, proposed another possibility. A universe that is curved—in a hard-to-imagine four dimensions— can be both finite and without boundaries. The surface of a sphere again serves as a two-dimensional analogy. Observers limited to the surface of the sphere and unable to perceive the third, vertical dimension could travel anywhere on the surface and never encounter an edge; they might even return to their starting place from the opposite direction. The surface of the sphere has no boundary and no center, yet it is finite; the same could be true of four-dimensional space-time.

But when Einstein got down to working out solutions to the field equations that would describe the actual universe, he began to travel what he described as a "rather rough and winding road." As far as anyone could tell from observations, the cosmos was a static entity: Its moving parts neither flew apart nor crashed together, and its size did not discernibly change with time. Yet no solution to the field equations produced a static model of the universe. Instead, all of Einstein's calculations indicated that the universe must either expand or contract.

In a very uncharacteristic step, Einstein did not pursue the implications of this finding. So convinced was he of the static nature of the cosmos that he decided to modify the equations, adding to them a term that corresponded to a cosmic repulsive force, working against gravity. The extra term, which he called the cosmological constant, seemed to make the problem of describing the universe more manageable. Because the constant was directly related to the size and mass of the universe, Einstein thought it would even be possible to determine those values from astronomical observations—although he never undertook this computational task himself.

Einstein seemed to have developed a complete and apparently unique mathematical description of the cosmos. However, Willem de Sitter, who had introduced Einstein's theory to Eddington, showed that another solution of the field equations was possible. De Sitter's model, which also incorporated a cosmological constant, was a mathematical description of a completely empty universe. A cosmos devoid of matter might seem absurd at first glance, but it is actually a pretty fair approximation of the real thing. Space is, after all, mostly empty.

Einstein and de Sitter were soon joined in the cosmological arena by Alexander Friedmann, a versatile Russian scientist who had originally made a name for himself in meteorology and related fields. When World War I broke out, Friedmann volunteered for service in an aviation detachment, where he helped establish meteorological and navigational services. In the summer of 1917, he became director of the first factory in Russia for making aircraft instruments. In 1918 he returned to teaching and research as a professor at the Academy of Sciences in what is now Leningrad.

Intrigued by Einstein's work in gravitation and cosmology and by the mathematical challenge of the field equations, Friedmann set out to find as many solutions as possible, without regard to their consequences for the real cosmos. Friedmann showed that the equations allowed for a wide variety of universes. In particular, he found that if he dropped the cosmological constant, all the results were expanding, matter-filled universes. Friedmann's solutions could be divided into two classes: those in which the universe would expand forever, and those in which the gravitational attraction of matter would eventually overcome the expansion, causing an ultimate collapse.

The factor that tips the balance one way or the other between expansion and collapse is the average density of mass in the universe. If the average amount of mass in a given volume of space is less than a critical value (which Friedmann, concerned only with theory, did not calculate), the universe will expand forever. Spacetime in such a universe is said to have negative curvature—analogous to a concave curve in ordinary space. Furthermore, such a universe is infinite, but the problem of precisely balancing the distribution of matter in an infinite universe governed by Newtonian gravity does not arise. General relativity dictates that gravity, like everything else in the universe, is limited by the speed of light; it cannot, as Newton assumed, act instantaneously over any distance. Thus, gravita-

Alexander Friedmann, a Russian mathematician at the Academy of Sciences in Petrograd (now Leningrad), published a paper in 1922 challenging Einstein's assertion that the universe was static. Discovering an error in Einstein's calculations, Friedmann showed that the cosmos would either expand indefinitely or eventually collapse, depending on its mass.

tional fields from infinitely large distances would take an infinite amount of time to make their influence felt, rather than all coming to bear at once on every bit of matter in the universe.

If the average density of mass is greater than the critical value, the universe will eventually collapse again into a dense concentration of matter, from which it might rebound to start a new cycle of expansion and collapse. Such a universe is an expanding version of Einstein's original static universe. It has positive curvature—analogous to a convex curve—and a finite radius, and it contains a finite amount of mass.

On the boundary between these two types of universes is one in which the average density of mass is equal to the critical density. Such a universe has zero curvature, and space-time is said to be flat because the usual Euclidean geometry for flat space applies. A flat universe is infinite and expands forever.

Friedmann's work was published in a well-known and widely read German physics journal in 1922. Einstein noticed the paper, disagreed with the results, and promptly published a paper of his own in refutation. Within a year, however, he had reconsidered. Friedmann's solutions to the field equations were mathematically correct, Einstein admitted. Nevertheless, those solutions seemed to Einstein to have no physical validity: To produce a curved universe with the apparently static characteristics observed by astronomers, something equivalent to the cosmological constant was still needed.

Neither Einstein nor Friedmann attempted to resolve their philosophical difference. The debate, conducted in terms of mathematics and not astronomy, never touched on the question of how expanding space-time might manifest itself in the night sky. Einstein was by this time beginning to work in other fields, and Friedmann, his curiosity satisfied, also turned away from relativity. Then, in 1925, at the age of thirty-seven, Friedmann died of pneumonia, which he contracted after becoming chilled during a meteorological balloon flight. Because the astronomical community had paid little attention to the debate, Friedmann's solutions to the field equations nearly died with him. Almost five years would pass before astronomers grasped the radical implications of relativity for their own work.

MEDITATIONS ON RELATIVITY

What if—? must have been a question Albert Einstein asked himself a dozen times a day. What if, for instance, you could ride a beam of light? Einstein posed that question when he was sixteen. Other sixteen-year-olds would have forgotten it the next day; Einstein came back to it years later, and his answers, in the form of the special and general theories of relativity, changed physics.

As illustrated on the following pages, many of the implications of relativity theory defy common sense and ordinary experience. On the whole, the seeming contradictions are a function of our poky existence: Humans do not zoom around at the speed of light. What Einstein discerned—largely through ingenious thought experiments—was that the laws of classical Newtonian physics, while adequate to explain the behavior of objects and forces in most circumstances, fall apart when the circumstances involve extremely strong gravitational fields or velocities approaching light-speed.

Most aspects of the physical world can be described in terms of three quantities—time, distance, and mass. Einstein showed that in all three cases, measured values will differ according to the location and motion of the measurer. One of the few constants in this relativistic universe is the source of Einstein's original inspiration, the speed of light itself.

WHEN OBSERVERS DISAGREE

The cornerstone of relativity is the idea that most measurements are not absolute. Instead, they depend on the motions of their measurers, most noticeably so at speeds close to that of light. Special relativity, a limited form of the theory illustrated here and on pages 42-47, describes the simplest case—how measurements are affected by uniform motion or travel at a steady speed in a straight line. Such motion is relative. To a picnicker in a grassy field, for example, a passing bus is in uniform motion; to a passenger on the bus, picnicker and field are in uniform motion headed the other way. In physicists' jargon, the bus and the field represent two distinct "frames of reference." The first assumption of special relativity is that observers inside a uniformly moving frame will find physical events within their frame unaffected by its motion. Moreover, as long as passengers on a smoothly moving vehicle have no external clues (such as a view of the passing landscape), they will be unable to detect that they are moving at all.

When observers look outside their frame, however, its motion relative to another frame affects what they see. The difference in perception is the direct result of special relativity's second assumption: that the speed of light, designated c, is constant for all uniformly moving observers. In addition, because light has a given speed—about 186,000 miles per second—a certain amount of time must elapse for light to travel from an event to an observer. Thus, observers moving at different rates relative to a light source will receive the light, and record the information it carries, at different times. One outcome is that inhabitants of different frames of reference will not agree whether given events are simultaneous *(right)*.

With the loss of simultaneity, the very concept of measurement is suspect. By definition, measurement, or comparison against a standard, must occur all at one time—for example, matching both ends of a piece of wood to a ruler simultaneously. But if simultaneity no longer applies, supposedly absolute values for length, time, and even mass enter a kind of limbo, in which different values are equally correct, depending on the observer's frame of reference.

At right, two hypothetical observers, one moving and one stationary, probe the meaning of simultaneity. Just as the traveling observer draws even with his fixed counterpart, lights placed at equal distances to either side of the two observers snap on, emitting beams that will very rapidly—but not instantaneously—reach the two experimenters.

A split second later, the moving observer—who, for the sake of demonstration, is traveling at an improbably high velocity near the speed of light—has progressed forward enough to encounter one of the beams. Thus, while the stationary observer still perceives both lights as off, to his moving colleague one light has turned on.

Suddenly the stationary observer sees both lights go on—at precisely the same instant. Because he and the two lights are all in the same, motionless frame of reference, the lights' synchronization is preserved. The other observer, in a frame of reference moving away from the rearward light, still believes that light has not come on.

At last, light from the lefthand source catches up to the moving passenger. Both observers see light from both sources now, but they disagree as to whether the lights came on simultaneously or sequentially. Neither observer can be considered right or wrong. Each is right from his point of view.

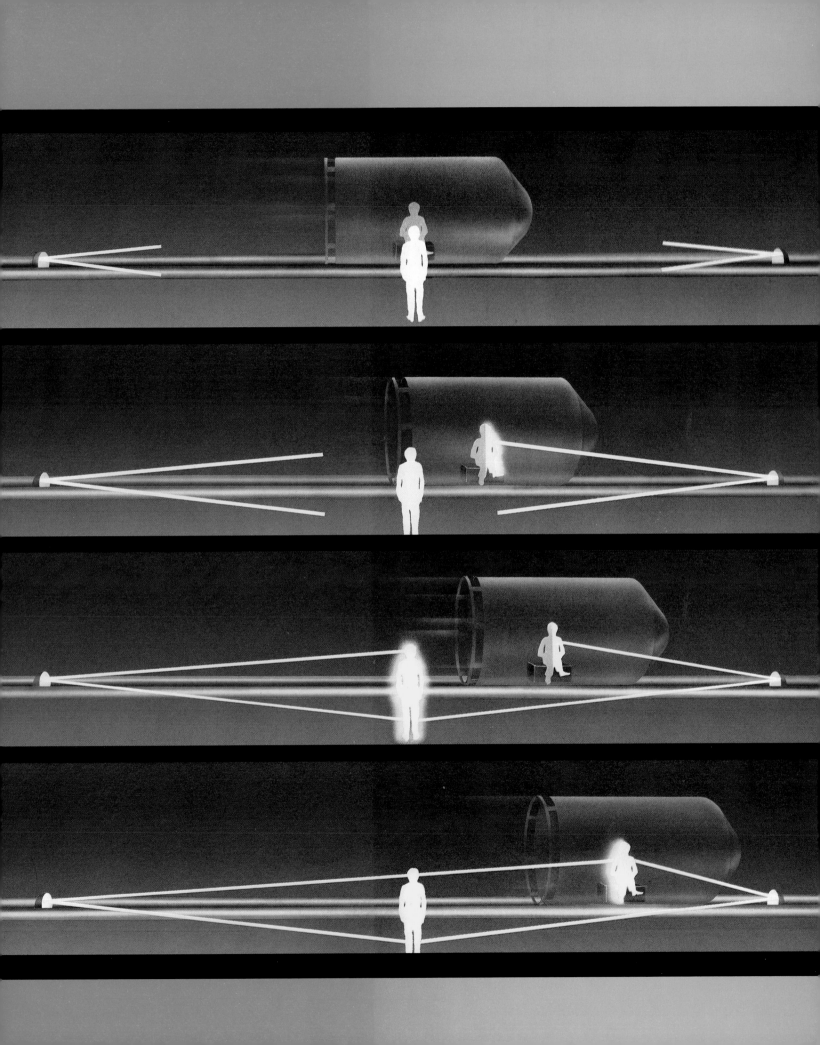

THE CURIOUS ELASTICITY OF TIME

Simultaneity is not the only aspect of time affected by relative motion. From certain perspectives, time itself will literally pass more slowly, as though it has been stretched out, or dilated. As with other relativistic effects, this so-called time dilation becomes noticeable only at very high velocities. But the phenomenon is real, demonstrated not just in thought experiments like the one below but also in physical observations.

For example, scientists note that unstable subatomic particles called muons, when observed at very high speeds, last nine times longer before decaying than do muons at rest. Employing a pair of equations known as the Lorentz transformations—which can numerically relate time and distance measurements made in different frames of reference—researchers have shown that the extended life span of moving muons can be precisely accounted for by applying relativistic stretching to their extended time scales.

A decaying muon, like a watch or a hypothetical light clock *(below)*, represents a system of predictable events. Because any such system tracks the passage of time, it may be regarded as a clock. Since light from a clock, whether reflected or directly emitted, can move

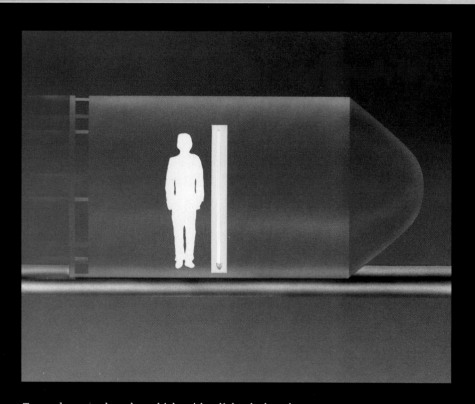

To an observer aboard a vehicle with a light clock—a hypothetical device that bounces light from a lower to an upper mirror—the clock pulses at the same rate whether the vehicle is moving or at rest. By dividing the height of the clock by the speed of light—an absolute value—the passenger can determine the duration of each light pulse. But, as shown at right, an external observer, in a different frame of reference, will obtain a very different duration measure.

toward observers only at the absolute speed of light, the clock's own travel toward or away from an observer will tend to alter the observer's perception of the clock's processes. In the comparison below, a clock moving away slows down; a clock moving close to the speed of light will very nearly stop.

Perhaps the oddest aspect of time dilation is that it works both ways, since each of two frames of reference is in motion relative to the other. Just as a stationary observer would see a moving clock as delayed, a moving observer would perceive that a stationary clock was running slow. If the observers are viewed as biological clocks, each will perceive the other to age more slowly. That both can be correct defies common

sense. The key to the paradox is that special relativity applies only to cases involving uniform motion. As long as both observers are moving steadily, they cannot be brought together to compare their clocks—or their ages. Doing so would require changing the direction and speed of travel of at least one observer, which would in turn break the uniform motion assumption.

Once outside the domain of uniform motion, special relativity no longer applies. The rerouted observer would be in accelerated motion (pages 48-51), and different assumptions would take over; whichever observer remained in uniform motion at the moment of comparison would be the older.

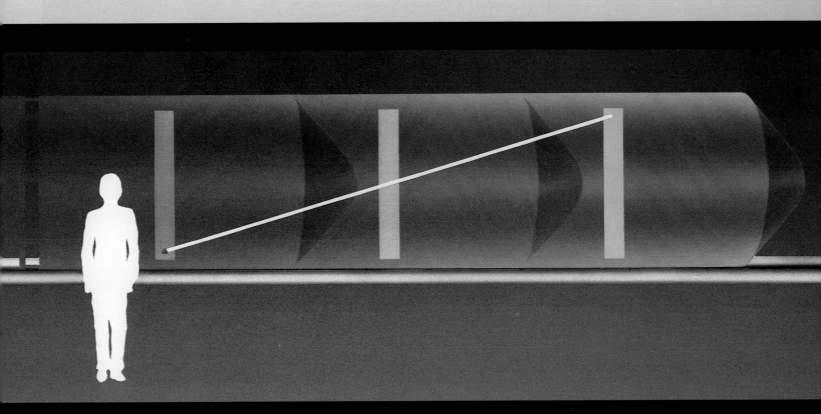

From a stationary observer's perspective, the moving vehicle's light clock runs slow. Movement causes each light pulse to travel in the direction of motion as well as upward to the top of the clock, a combination that results in a long diagonal course (yellow). Since the speed of light is fixed, the light pulse, of necessity, takes longer to reach the end of its longer path. The faster the vehicle moves, the longer the diagonal and the slower the pulse appears to the stationary observer.

THE VARIABILITY OF LENGTH IN MOTION

In the strange world of special relativity, length and distance, like time, vary according to the relative motions of the observer and the observed. Any object moving at a very high, constant speed will appear to relatively stationary observers to contract in the direction of its motion. (Height remains unchanged, however.) As with time dilation, the Lorentz transformations describe the precise numerical relationship between speed and length.

The effect of motion increases drastically at very high speeds. For example, a vehicle traveling at half the speed of light will appear to stationary outside

Two observers try to measure a vehicle's length by comparing it to the known length of a stationary tunnel. Lights at each end of the tunnel will let the observers make the comparison no matter how fast the vehicle runs. The light on the right will turn on when the front of the vehicle emerges from the tunnel; the light on the left will flash on as the back of it enters. As the experiment proceeds, the observer across from the tunnel's midpoint sees both lights come on simultaneously *(below)* and concludes that vehicle and tunnel match exactly *(left)*.

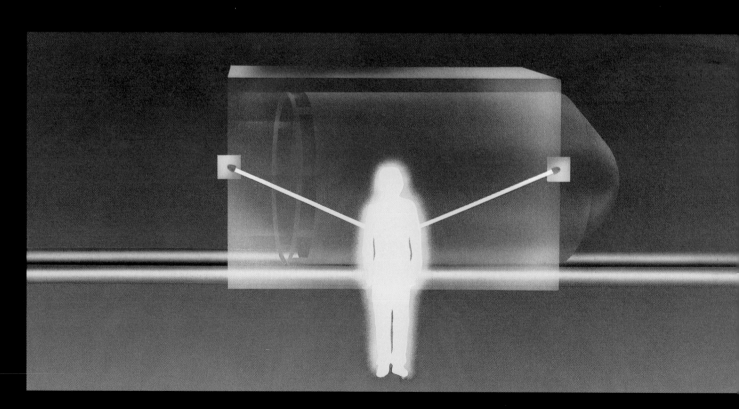

spectators to be about seven-eighths the length its passengers measure; a vehicle reaching 90 percent of light-speed will seem to be less than half as long as its rest length.

These measurements vary because the speed of light does not—confounding all attempts to arrive at measurements common to frames of reference that are moving relative to one another. In the hypothetical thought experiment below, observers in two different frames try to determine the length of a trainlike vehicle by using as their yardstick a tunnel whose length is known. Light signals at either end of the tunnel allow both of the observers to compare vehicle and tunnel.

Like time dilation, length contraction is reciprocal. Each of two relatively moving observers will see the other's surroundings contract in length, while his or her own environment seems unaffected. In the example below, the fixed observer thinks a moving vehicle is shorter than the length ascribed to it by the on-board passenger, yet both are correct. Again, the paradox occurs because the surroundings of the two observers can never truly be compared without matching their speeds—a change that would break the bounds of uniform motion imposed by special relativity.

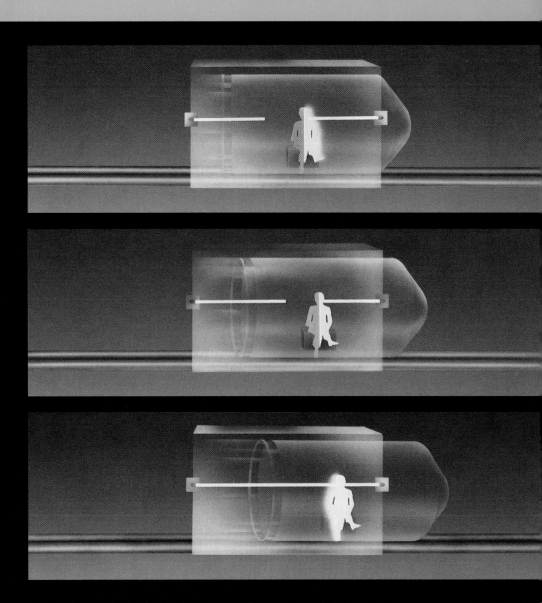

To the on-board passenger, the lights tell a very different story. As in the simultaneity experiment (pages 40-41), the motion of the vehicle carries the passenger forward to meet the light emanating from the forward signal. Receiving that light before any light arrives from the signal behind him, he concludes that the front end of the vehicle has left the tunnel before its back end has entered and therefore that the vehicle is longer than the tunnel.

The front of the vehicle has emerged from the tunnel, but the passenger has yet to see the rear light turn on. His estimate of the vehicle's length increases; the growing pause between signals is an indication of how much longer it is than the tunnel.

At last, the passenger sees the second light flash on, signaling that the back of the vehicle has entered the tunnel. On the basis of the interval between the two signals, he decides that the vehicle is longer than the tunnel through which it passed—a conclusion at odds with that of the stationary observer.

APPROACHING THE ULTIMATE SPEED LIMIT

According to the laws of classic Newtonian physics, the steady application of a great enough force could propel an object to any speed—theoretically, even to one greater than that of light. As he contemplated this possibility, Einstein recognized that it would have unacceptable consequences for cause-and-effect relationships. For example, a tennis ball returned at faster than light-speed could arrive before it was served. To solve this paradox and not violate the laws of classical physics, he theorized that, like time and distance, an object's mass—its resistance to a change in motion—must vary with its motion.

According to Einstein's special relativity, then, the mass of an object increases along with its speed relative to an observer, requiring ever more energy to make it move faster. To move an object at the speed of light would, by definition, require an infinite amount of time and force—an obvious impossibility. Therefore, not only is the speed of light—around 186,000 miles per second—a constant that remains absolute regardless of an observer's frame of reference, it also represents an ultimate speed limit.

Other meditations on mass yielded Einstein's most famous equation, $E = mc^2$, in which E stands for energy, m for mass, and c for the speed of light. What Einstein saw was that any increase in an object's motion, or kinetic energy, effectively translates into an increase in its mass. Mass and energy must therefore be equivalent—an insight that would have enormous implications for later work on quantum mechanics and theories of the early universe.

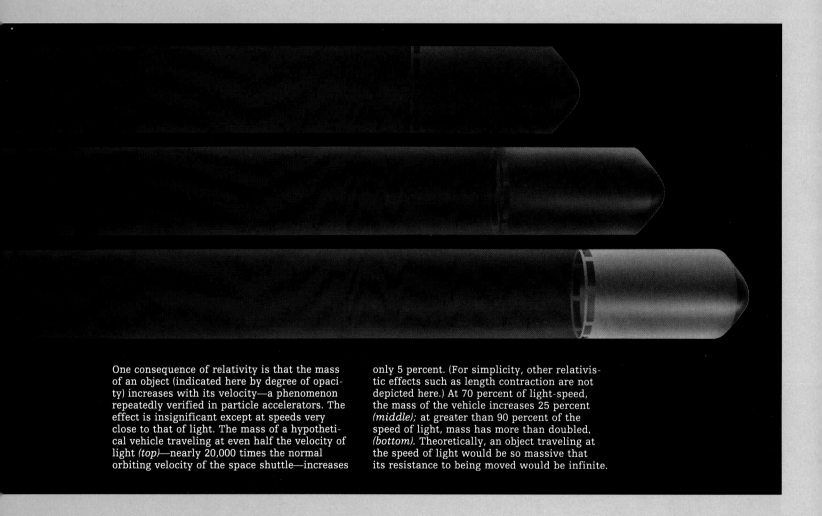

One consequence of relativity is that the mass of an object (indicated here by degree of opacity) increases with its velocity—a phenomenon repeatedly verified in particle accelerators. The effect is insignificant except at speeds very close to that of light. The mass of a hypothetical vehicle traveling at even half the velocity of light *(top)*—nearly 20,000 times the normal orbiting velocity of the space shuttle—increases only 5 percent. (For simplicity, other relativistic effects such as length contraction are not depicted here.) At 70 percent of light-speed, the mass of the vehicle increases 25 percent *(middle)*; at greater than 90 percent of the speed of light, mass has more than doubled. *(bottom).* Theoretically, an object traveling at the speed of light would be so massive that its resistance to being moved would be infinite.

INTO THE FOURTH DIMENSION

A fundamental implication of the theory of special relativity is that time and space are intimately intertwined, despite the human tendency to treat them as separate and unrelated. Motion, for example, is a change in spatial position over time; the speed of light is measured in distance traveled over time; an event is specified not only by when it occurs but where it occurs. For relativity theorists, then, the universe is conceived as a four-dimensional space-time continuum in which time is a coordinate as crucial as length, width, or height.

In space-time terms, a brick is not simply an object located at a point in a wall; rather, it is a space-time "event" that exists at a particular place at a particular time. Scientists groping for a term to encompass all such events settled on "the world," and they named the coordinates describing an event's successive space-time locations its "worldline."

The notion of space-time is a convenient mathematical artifice because calculations based on its four dimensions allow physicists to reconcile the disparity between the perceptions of observers moving relative to one another. No matter what their individual assessments of length, simultaneity, or duration of events within their own frames of reference, all observers must obtain the same value for the distance traveled by light in a given time. They can thus agree on the "space-time interval" between two events—a mathematical expression that encompasses both time and space in a single figure.

Illustrated at right is a three-dimensional diagraming technique, devised by the Russian-born mathematician Hermann Minkowski, to help visualize four-dimensional worldlines.

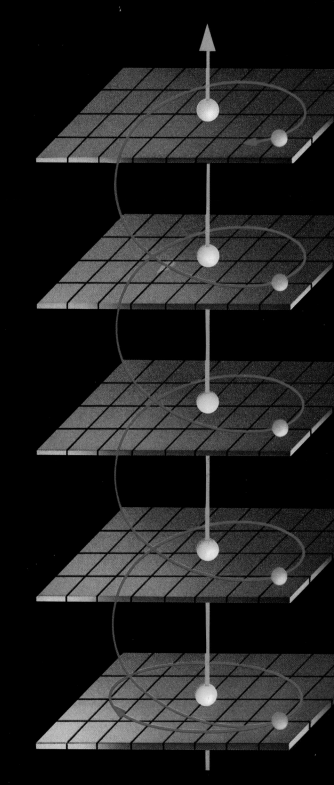

Above, a smoothly orbiting planet traces its worldline through a modified Minkowski diagram of space-time. In each plane, grids denote the planet's position relative to the star at center, incorporating two of the three spatial dimensions. The third dimension—height above or below the star—cannot be shown in the diagram, for here the vertical axis represents time. Stacked planes indicate successive years as the planet orbits through the same spatial location again and again.

THE PRINCIPLE
OF EQUIVALENCE

In the real world, smooth, uniform motion is more the exception than the rule. Technically, any change in speed or direction is called acceleration, which can thus mean slowing down as well as speeding up, or simply a redirection. Ordinarily, an observer in an accelerating frame of reference can perceive its motion. Passengers in an automobile, for example, feel themselves pressed backward if the car starts suddenly from a dead stop. Their awareness seems to imply that acceleration is absolute, not relative; they need not refer to anything outside their frame to detect their own motion. But if accelerated motion is absolute, it would have to be subject to a different set of natural laws from those that apply to uniform motion—a proposition that Einstein found highly objectionable. He thus set out to devise a more general theory that would apply to motion of all sorts. In the process, he developed a new theory of gravity.

The starting point was Galileo's finding that falling objects accelerate at the same rate despite differences in their mass: If dropped from the same height in a vacuum, a cannonball and a pea will hit the ground at the same time. Einstein was skeptical of Newton's explanation that the force of gravitational attraction precisely equaled an object's inertial mass. He rejected the notion that this uncanny coincidence was merely an accident of nature.

Then came the sweeping insight now known as the principle of equivalence. As depicted in the thought experiment at right, Einstein imagined an observer in deep space who would not be able to tell whether he was on the surface of a planet—and subject to its gravity—or in accelerated motion at a rate that precisely replicated the force of gravity on the planet. If the effects of acceleration and gravity are indistinguishable, then acceleration is not absolute after all. And if gravity's effects are equivalent to those of acceleration, then the mystery of Galileo's falling bodies is solved: Objects of differing mass fall equally under gravity because they are behaving exactly as they would if they were in a vehicle in free space and the vehicle was accelerating toward them.

Adrift in space? A passenger in the closed vehicle at right could explain his sensations and the behavior of objects in the vehicle with him strictly in terms of acceleration, by hypothesizing it to be a ship traveling through deep space. When he finds himself floating weightlessly *(near right)*, it could mean the spaceship is traveling at a constant speed; when he feels pressed firmly toward a surface that suddenly becomes "down" *(far right)*, it may signify that the ship's speed is increasing. If he dropped two balls under these conditions, their behavior would be perfectly consistent with the spaceship theory, since the floor, in effect, accelerates up to meet them.

Or falling earthward? With equal validity, the passenger in the hypothetical vehicle could speculate that he is in a sealed elevator, affected only by Earth's gravity. In the scene at near right, floating contents could indicate that the elevator's cable has snapped and the car is falling freely under gravity's pull. At far right, firm footing and a sensation of weight could imply the car is simply at rest. Two dropped balls behave as they do on Earth, hitting the floor simultaneously. That the passenger could explain his situation using rules that apply either to gravity or to acceleration demonstrates that the two are equivalent. (The principle of equivalence is true in local regions of space; at large scales, gravity would make the paths of objects converge toward the center of the attracting body.)

HOW GRAVITY HINDERS TIME

If accelerated motion is relative, all space and time are up for grabs. Just as uniform motion at speeds approaching that of light affects such measurements as the length of objects or the duration of events, so does accelerated motion—and so, by the principle of equivalence, does gravity. The thought experiment at right demonstrates that time slows as a result of acceleration. Actual experiments have repeatedly confirmed that gravity produces the same effect: A clock near a massive body does run more slowly than one farther away.

In the 1960s, for instance, researchers found that clocks in the mile-high city of Boulder, Colorado, gained about fifteen-billionths of a second per day compared with clocks located near sea level in Washington, D.C. They attributed the difference to Boulder's greater distance from the Earth's gravitational center. The time dilation effect is even noticeable on a much smaller scale. Other experimenters have shown that, remarkable as it seems, a clock on the ground floor of a building seventy-four feet tall runs more slowly than one near its top.

The mind-stretching implications of these phenomena have long served as a launching pad for science fiction: The secret to long outliving one's earthbound children—or great-great-grandchildren—would seem to be to stow away on a spaceship accelerating to near light-speed. (Failing that, simply live in a bungalow at sea level to outlive people who favor penthouses in the Swiss Alps.)

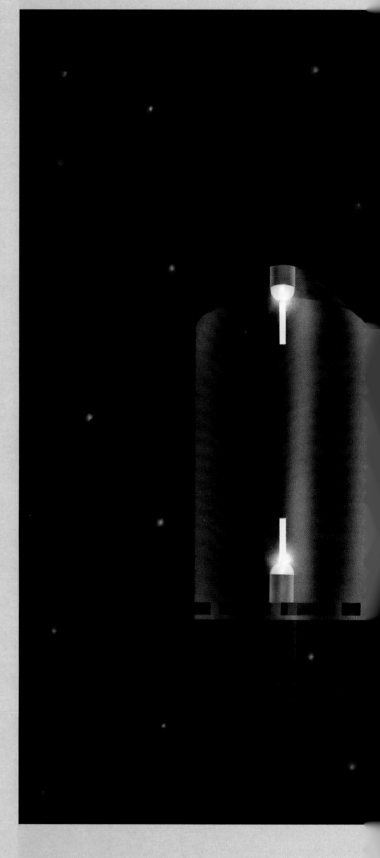

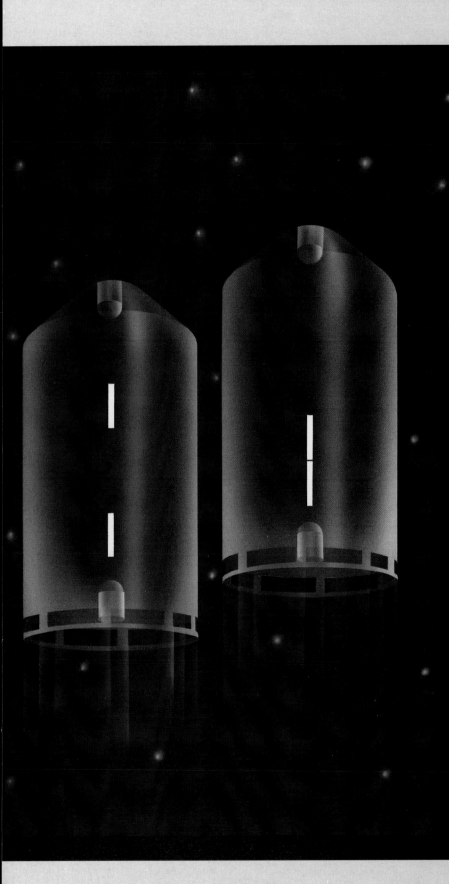

In a study of acceleration's effect on time, atomic clocks at opposite ends of an accelerating spaceship emit flashes of light directed toward the ship's center. Although the flashes seem synchronized at first, the ship's acceleration over the next instant prevents them from reaching the midpoint simultaneously. Instead, the lower clock's pulse appears to lag behind. An outside observer would say the pulse was delayed because the midpoint of the ship accelerated away from it, lengthening its path. To an observer inside the ship, however, the lower clock would simply be slow.

CURVING THE CONTOURS OF SPACE

Another bizarre aspect of relativity theory is that acceleration and gravity, in addition to affecting the measure of time, do strange things to the dimensions and geometry of space. As shown here, the shortest distance between two points in an accelerated area may turn out not to be a straight line—provided the observer is also in the accelerating frame of reference.

Gravity's effect is identical. Since gravity is a property of mass, scientists generally look for spatial distortions near massive bodies such as stars or even galaxies. Warped geometry is exceedingly difficult to

detect directly, but astronomers can observe it indirectly by monitoring such phenomena as the passage of light in suspect regions. Under ordinary conditions, light travels in a straight line, the shortest path between two points. Any indication that light has followed a curved trail suggests that, for some reason, the shortest path is no longer straight; space itself must be deformed in that area.

An early confirmation of general relativity theory was a successful attempt in 1919 to detect curved starlight by taking advantage of a solar eclipse. Stars near the edge of the darkened Sun were photographed to note their positions. These positions were then compared with the stars' positions as determined normally—at night, when the Sun was not in the way. During the eclipse, the stars appeared to have shifted from their normal locations because the path of their light was warped as it passed close to the massive body of the Sun on the way to Earth. Even more extreme warping is caused by so-called gravitational lenses, very massive bodies that deflect passing light in such fashion as to create multiple images *(pages 112-115)*.

As seen at left from an external, relatively stationary perspective, a beam of light that originates outside a passing spaceship follows a straight path from left to right through the ship's glass walls. Because the hypothetical ship is moving at near light-speed itself, and because the light beam requires a certain amount of time to travel the width of the ship, the beam enters near the top but hits the opposite wall at the bottom. The rapidity with which the light beam seems to drop from one instant to the next indicates that the ship is steadily accelerating upward.

Aboard the ship *(right)*, an observer sees the light travel along a curved track, one that sums together each of the intermediate points shown at left. Because light always takes the shortest path possible, the observer concludes that the geometry of the ship's internal space has been warped by its acceleration until arcs are shorter than straight lines. Since acceleration and gravity are equivalent under relativity theory, gravity too should alter the geometry of space *(pages 54-55)*.

BUMPS AND DIPS IN THE COSMIC FABRIC

Once astronomers and cosmologists recognized that gravity profoundly affects both space and time, they began referring to gravity as one of four-dimensional space-time's geometrical properties. A useful two-dimensional analogy is to think of space-time as a taut, infinitely stretchable rubber sheet. Where there is little significant mass—that of a small asteroid, for instance—the sheet is nearly flat. But where massive bodies such as planets and stars press into the sheet from above or below, they deform it into localized bumps and dips, sometimes called gravity wells.

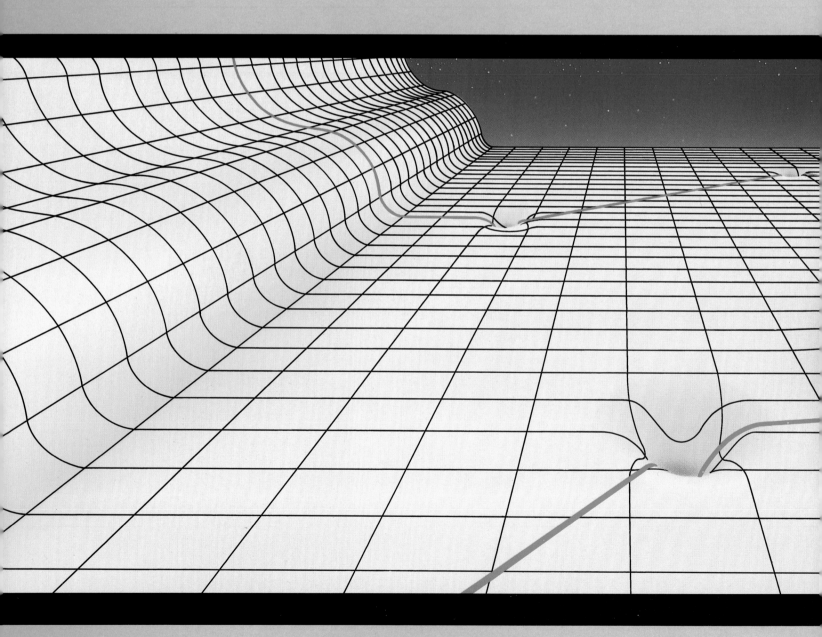

Objects traveling along the sheet follow the deformations. A comet moving toward a star, for example, will sink into the star's gravity well. The descent will deflect the comet's path, perhaps enough to trap it into circling the sides of the well indefinitely. From general relativity's perspective, moving bodies like the comet are not pulled off course by gravity; they are following the mass-warped contours of space-time itself.

A rigorous description of such wells and swells requires geometry systems that contradict the principles developed by the Greek geometer Euclid in the fourth century BC. Euclidean geometry applies to flat surfaces; for curved space, other rules apply *(pages 56-57)*. Navigators have long used one non-Euclidean system, spherical geometry, to plot the shortest routes between points on the curved surface of the Earth. The geometries are equally effective in four-dimensional space-time, yielding equations that chart exactly how space-time will curve in the presence of a given mass. Using such mathematical tools, astronomers can model the gravitational effects of all manner of phenomena, from black holes to double stars.

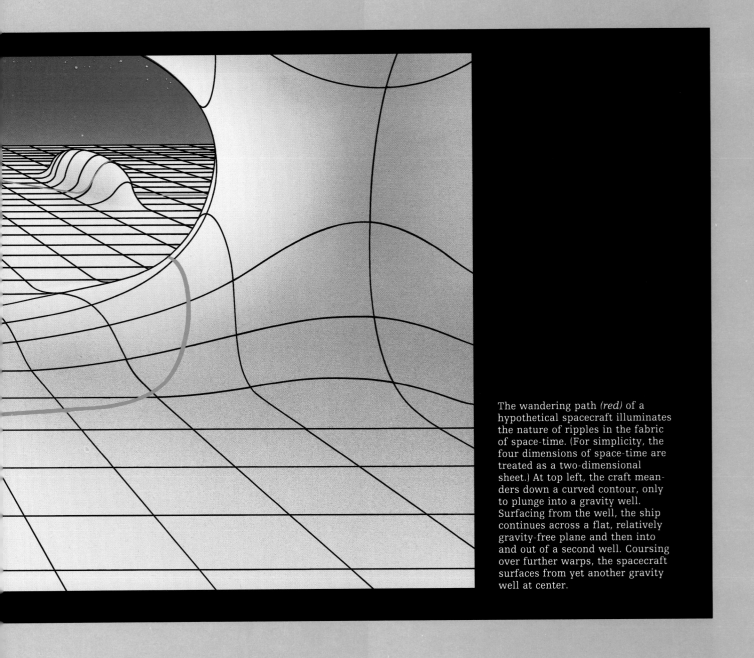

The wandering path *(red)* of a hypothetical spacecraft illuminates the nature of ripples in the fabric of space-time. (For simplicity, the four dimensions of space-time are treated as a two-dimensional sheet.) At top left, the craft meanders down a curved contour, only to plunge into a gravity well. Surfacing from the well, the ship continues across a flat, relatively gravity-free plane and then into and out of a second well. Coursing over further warps, the spacecraft surfaces from yet another gravity well at center.

THE FATE OF THE UNIVERSE

One consequence of general relativity's description of gravity and its relationship to space-time is that it allows cosmologists to visualize possible answers to a double-barreled question that until recently was the province of religion, mysticism, and philosophy: Will the universe end, and if so, how?

The possible answers take the form of three cosmic models *(right)*. All begin with the assumption that the universe is expanding, probably from the cosmic explosion known as the Big Bang. Deciding which model is the right one depends on finding a small but important number: the average density of matter in the universe. Average mass density, as it is known, determines whether or not gravity will ultimately act to stop expansion.

If the average mass density is below a critical value of about three hydrogen atoms per cubic meter, the four-dimensional universe of space-time is "open" *(far right)*. It has a shape difficult to think of in three dimensions—much less represent in two—and it will continue to expand forever, until every atom is so far away from every other atom that a nearly energyless Big Chill pervades. If the average density is above the critical value, the cosmos is closed *(center)*. Eventually, the gravitational attraction of all the mass in the universe will halt expansion and draw the cosmos back in on itself until it implodes in a Big Crunch. Finally, if the average density is equal to the critical value, then the universe is essentially flat *(right)*; gravity will slow expansion but never quite stop it.

For theoretical and observational reasons, some cosmologists believe that the amount of matter in the universe exactly equals the critical density and that the universe is flat. However, direct observation finds a density ten to a hundred times less than the critical value, in which case the universe is wide open. Scientists are increasingly suspicious that the reason they cannot find more matter in the universe is that most of it simply does not radiate in a form that can be detected with current technology. Unless the average mass density of the universe is somehow ascertained, the fate of the cosmos will remain a hotly debated question.

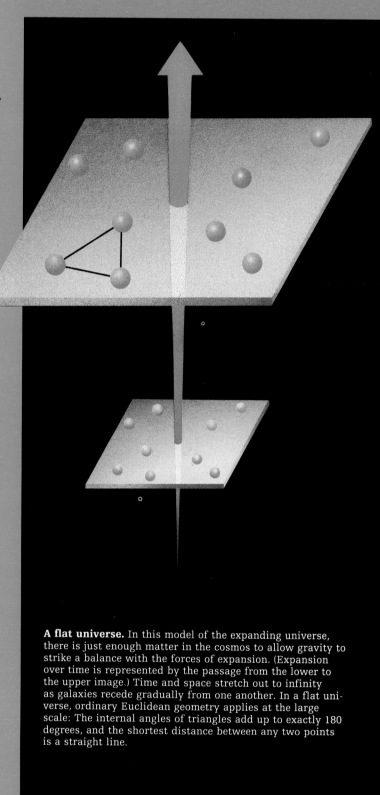

A flat universe. In this model of the expanding universe, there is just enough matter in the cosmos to allow gravity to strike a balance with the forces of expansion. (Expansion over time is represented by the passage from the lower to the upper image.) Time and space stretch out to infinity as galaxies recede gradually from one another. In a flat universe, ordinary Euclidean geometry applies at the large scale: The internal angles of triangles add up to exactly 180 degrees, and the shortest distance between any two points is a straight line.

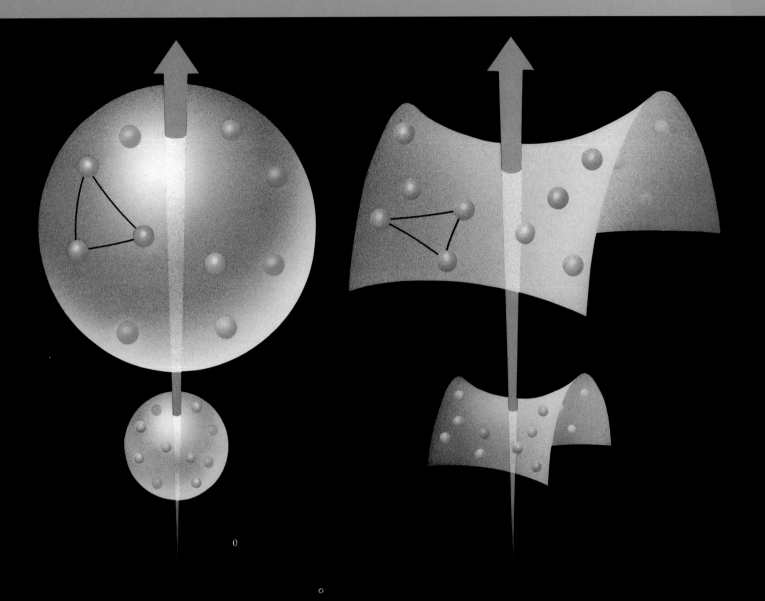

A closed universe. A universe with an average density of matter that exceeds three hydrogen atoms per cubic meter is, like the finite surface of a sphere, finite in time and space but without boundaries. In a closed universe, gravity will ultimately cause it to begin contracting, until all matter is concentrated in an infinitely dense state. Here, a non-Euclidean spherical geometry applies: Just as on the surface of a ball, the sum of a triangle's internal angles is always more than 180 degrees, and an arc is the shortest distance between two points.

An open universe. Because the average density of matter in an open universe is below the critical value of three hydrogen atoms per cubic meter, gravitational forces cannot act to stop expansion. The resulting configuration of the cosmos is represented here by a saddle shape known as a hyperboloid. Like the flat universe, the open universe is infinite in both time and space; however, its geometry is very different from that of its theoretical cousin. Here, the internal angles of a triangle total less than 180 degrees, and the shortest path between any two points is an arc.

2/Redefining Reality

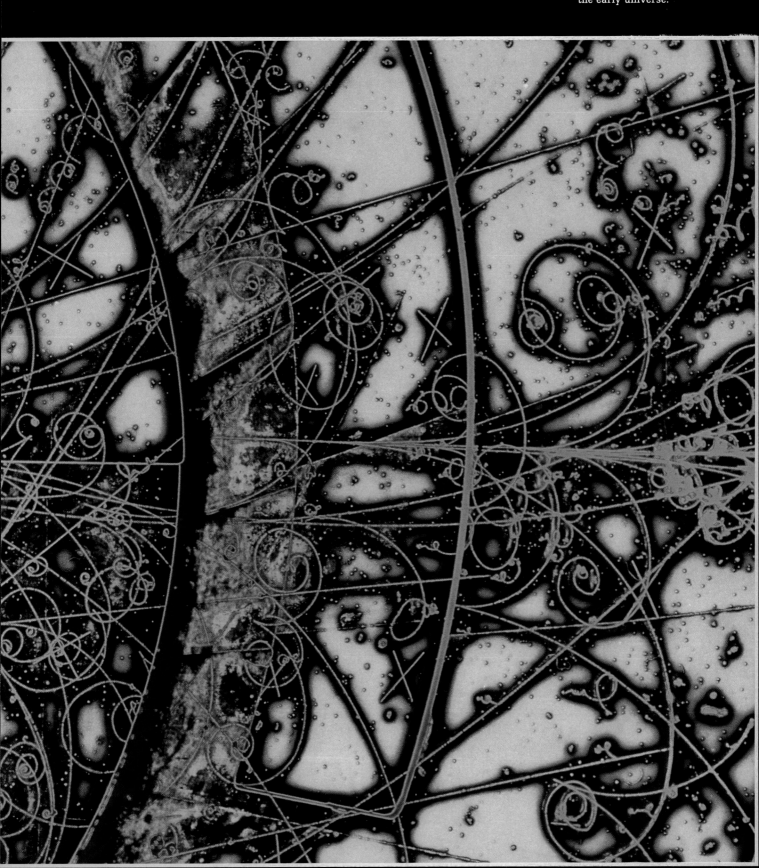

hydrogen, a proton struck by an invisible speeding neutrino disintegrates into a spray of subatomic particles in a reaction that approximates conditions that prevailed in the early universe.

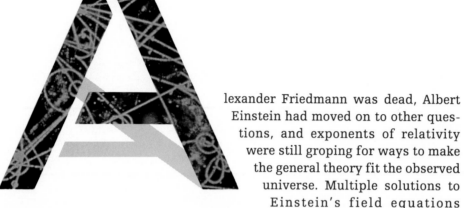

lexander Friedmann was dead, Albert Einstein had moved on to other questions, and exponents of relativity were still groping for ways to make the general theory fit the observed universe. Multiple solutions to Einstein's field equations seemed to allow universes that were mutually exclusive—on the one hand, static, on the other, expanding, and either empty or filled with matter. None of the solutions satisfactorily described the cosmos as it appeared, both static and sparsely populated with stars and galaxies. Before such a description was found, physics would undergo a number of revolutions and counterrevolutions, sparked by successive insights into the nature of the atom and of energy. A new language would evolve—quantum mechanics—whose bizarre grammar would ultimately lead theorists to the primeval moment: the beginning of the universe itself.

Fittingly, one of the first to dare describe creation in nonbiblical terms was himself a priest as well as a physicist. Raised in Charleroi, Belgium, Georges Lemaître was educated at a Jesuit school and went on to study civil engineering at the Catholic University of Louvain. At the outbreak of World War I, when he was scarcely twenty, he volunteered for the army, earning decorations for his service as an artillery officer. After the war he returned to the university but changed his focus from engineering to mathematics and physics, with a leavening of philosophy courses. Shortly after graduating in 1920, the war veteran entered a seminary. Lemaître was an unusually independent thinker, and he explored scientific matters—particularly the new doctrine of relativity—even as he pursued his theological studies at the seminary. By 1923 he was both an ordained priest and a promising theoretical physicist. For the new Father Lemaître, these twin callings were not as irreconcilable as they seemed. To his mind, they not only involved different mental processes but also led to different understandings of the world.

It was as a student of physics that he embarked in 1923 on a two-year journey that took him from the scientific backwater of his native Belgium to centers of advanced research in England and the United States. He first stopped at Cambridge to study for a year under Arthur Eddington, then director of the Cambridge Observatories. Eddington inspired the young Belgian to specialize in general relativity, and wrote a lengthy introduction for

Lemaître's first research paper on the subject. In correspondence with a colleague, he lavished praise on the mathematical ability of his "wonderfully quick and clear-sighted" student.

In the summer of 1924, Lemaître moved on to the United States, where he worked at the Harvard Observatory while preparing for a Ph.D. in astronomy at the Massachusetts Institute of Technology. The Belgian took full advantage of his proximity to prominent American astronomers, seeking them out to discuss the latest observations and cosmological theories. He was especially intrigued by indications that extragalactic nebulae, as other galaxies were then known, were receding from the Milky Way at speeds in the neighborhood of 700 miles per second. In his first paper on relativistic cosmology, published in 1925, Lemaître's careful mathematical analysis revealed a new property of the model of an empty universe proposed by Willem de Sitter in 1917. De Sitter's universe was nonstatic, meaning its dimensions changed with time. Lemaître noted that this property might explain the observed recession of extragalactic nebulae. They would be carried apart by the stretching of space-time. In the end, however, Lemaître rejected the de Sitter model because it proposed a space with no curvature, clearly a violation of the principles of general relativity.

Decorated later by the Vatican for his scientific achievements, Belgian priest Georges Lemaître galvanized cosmologists in 1927 with his proposal that a hot, dense "primeval atom" exploded to create the present universe. Lemaître's revolutionary concept prefigured the now-accepted Big Bang theory of the origins of the cosmos.

By mid-1925 Lemaître was back at Louvain, where he would remain a member of the faculty for nearly four decades. He was a devoted teacher who especially enjoyed the noisy company of students, but his quickness of mind made attempts at scientific collaboration onerous, and he generally avoided it. A penchant for laboring in solitude was no handicap, however. In 1927 Lemaître made a conceptual breakthrough when he again pondered the behavior of extragalactic nebulae. (By this time these objects were generally recognized as galaxies in their own right.) Unlike his initial effort two years earlier, the paper he now published was not simply speculation; it featured a mathematical model of the universe that incorporated the concept of receding galaxies. Lemaître's solution to Einstein's field equations yielded a universe of constant mass, with a radius that, as he put it, "increases without limit" at a rate matching that of the receding star systems.

Although he did not know it, his calculations were a remarkable echo of those published by Alexander Friedmann in 1922. But where Friedmann had treated the subject as a theoretical exercise, Lemaître rigorously connected the mathematics to astronomical observations at each step. The resulting model was the first persuasive use of the principles of relativity to explain the real universe. Receding galaxies actually were flying apart, Lemaître said, borne along as the fabric of space-time itself spread.

The implications of Lemaître's work—the work itself, for that matter—went, like Friedmann's, essentially unnoticed for some time. Published in a relatively obscure Belgian scientific journal, his theory labored under the unwieldy title "A Homogeneous Universe of Constant Mass and Increasing Radius Accounting for the Radial Velocity of Extra-Galactic Nebulae." If other scientists saw it, they either did not read it or failed to understand it; the foremost theorists continued to gnaw at the problem of reconciling relativity with observations for more than two years. At a London meeting of the Royal Astronomical Society in early 1930, for instance, Eddington and de Sitter agreed that no adequate solutions to the field equations had yet come to light.

When Lemaître later read a report of that meeting, he wrote Eddington to point out the solution he had outlined in his 1927 paper. That letter proved to be a turning point in Lemaître's career. In its wake, both Eddington and de Sitter became champions of what Eddington called "Lemaître's brilliant solution." After Eddington arranged for an English translation of the paper—and Einstein himself followed up with an endorsement of the theory—Lemaître's scientific reputation was secure.

Even before his model received its belated due, Lemaître was moving on. Mentally running the film of an expanding universe backward, he envisioned the galaxies no longer carried apart but moving together. Space shrank. The distances between galaxies dwindled from the unimaginable to mere miles. Then, as the rest of the scientific world mulled the ramifications of his expanding-universe theory, the physicist-priest took the next step.

There might well have been, he wrote in 1931, a real beginning: Before expansion got under way, there existed a "primeval atom," with a weight equal to the total mass of the universe. Within its bounds, the forces of electrical repulsion were overcome, and matter existed in a highly compressed, exceedingly hot state. Lemaître's next speculation was an extraordinary intuitive leap: In a sort of superatomic decay, he said, this cosmic egg had flung its contents outward in a gargantuan explosion. "The last two thousand million years are slow evolution," he wrote. "They are ashes and smoke of bright but very rapid fireworks."

Lemaître's vision of cataclysmic birth did not sit well with one of his erstwhile supporters. Eddington considered the scenario "unaesthetically abrupt," preferring to think that cosmic change occurred in a more measured, elegant fashion. In his estimation, the universe had rested unperturbed for an infinite time and then, because of built-in instabilities, had slowly begun to expand. Eddington regarded observations of galactic recession as evidence that the universe had entered this second phase, which would end with a kind of cosmic sigh of exhaustion.

Neither Eddington nor Lemaître could muster arguments persuasive enough to sway the other. Lemaître, in developing his theory of the expanding universe, had relied on the principles of general relativity, a language of large masses and vast distances. Relativity allowed him to describe the expansion mathematically as beginning from a very small source. It did not allow him to penetrate the physical processes that would transform a tiny, dense clump

By demonstrating mathematically that atoms radiate energy in discrete packets, or quanta, the German physicist Max Planck engendered a new field of science—quantum physics—in 1900.

of matter into the observed universe of receding galaxies. What that task required was a mathematical language of the very small—a theory of nuclear structure that could be used to delineate the interactions that had taken place within the cosmic egg.

LIGHT FROM A BLACK BODY

The needed language and theory, now known as quantum mechanics, were even then evolving in the physics institutes of Europe. Formulation had begun at the turn of the century with the work of Max Planck, a German theoretical physicist who taught in Berlin. Planck was intrigued by a fundamental problem having to do with radiation from a so-called black body.

Scientists knew that the color of the light a body emits—its range of wavelengths—is related to the material the object is made of and to its temperature. Generally speaking, blue light, with short wavelengths, prevails in the spectra of very hot objects; red, or long, wavelengths indicate less heat. Other wavelengths are also represented, but as a rule, each temperature correlates with a dominant wavelength, giving a glowing object a characteristic color. To simplify their analysis of radiation, nineteenth-century theorists had conjured up the black body. Unlike real objects, this imaginary entity absorbs radiation of all frequencies—rendering it perfectly black. It also emits radiation of all frequencies, regardless of its material composition. Experimenters had created ingenious devices to approximate this theoretical construct in laboratories and had learned a great deal about the characteristics of black body radiation. What they lacked was a theory to predict the distribution or form of the black body radiation spectrum, that is, the amount of radiation given off at specific frequencies at various temperatures.

Most scientists believed that the key to this problem lay in understanding the interaction between electromagnetic radiation and matter. In 1900, when Planck attacked the problem, the accepted electromagnetic theory of light held that light was a wave phenomenon and that matter—assumed to contain small, electrically charged bodies, or particles—radiated energy in the form of light waves when these charged particles were accelerated. Accepted wisdom also decreed that the amount of energy radiated by an accelerated charged particle could fall anywhere along a continuous range.

For the purposes of studying black body radiation, Planck imagined charged particles as tiny oscillators, repeatedly accelerated and decelerated in a simple, smooth, regular fashion, as if they were attached to a weightless spring. So far, he was firmly within the realm of nineteenth-century physics. But then he made a radical departure.

EINSTEIN AND THE PHOTOELECTRIC EFFECT

Toward the end of the nineteenth century, scientists discovered that electrified sheets of metal exposed to light gave off charged particles, later identified as electrons. This behavior, which came to be known as the photoelectric effect, was not especially surprising: Earlier investigations by Scottish physicist James Clerk Maxwell and others had revealed light to be a wave that transported electric forces. It seemed plausible that such waves could shake electrons loose from their atomic moorings. But when physicists tried to examine the kinetic energy of the freed electrons, problems arose. By the logic of the wave theory, bright light should shake up the electrons the most, sending highly energetic particles zinging from the metal plate; electrons freed by the gentle push of dim light should have much less kinetic energy.

This reasoning was not borne out by experiment. In 1902 Philipp Lenard, a professor of physics at the University of Kiel, demonstrated that intensifying the brightness of the light hitting the metal yielded a greater number of cast-off electrons but did not affect their energy in the expected way. Apparently their energy depended not on the intensity but on the frequency of the incoming light: The higher the frequency, the peppier the emerging electrons. Low-frequency red light, no matter how bright, rarely ejected electrons at all, whereas high-frequency blue and ultraviolet light—no matter how dim—almost always did. According to standard physics, which viewed light as a wave phenomenon, the results did not make sense.

Albert Einstein came to the question a few years later in the course of examining Max Planck's theories on radiant energy. In a bold extension of Planck's work, Einstein proposed that light was made up of particles (now called photons) rather than waves. This, he said, would neatly explain the photoelectric effect. Each photon would contain a certain amount of energy; photons at higher frequencies would contain more than those at lower frequencies. Individual electrons in the metal plates would absorb the energy of individual photons. If that energy was high enough, the invigorated electron might fly free of the plate. In addition, because bright light contains more photons than dim light, it should shake loose more electrons. At any given frequency, the brighter the light, the denser the hail of photons and the greater the number of electrons freed.

Einstein's explanation of the photoelectric effect was published in 1905, the same year as his paper on special relativity, reopening an issue that most physicists had thought settled. His assumption of light-as-particle countered the light-as-wave theory, and the argument continued for nearly two more decades, until physicists accepted that, somehow, light is both. Even before the debate was resolved, however, Einstein's breakthrough work on the photoelectric effect earned him the most coveted award in physics, the Nobel prize.

In the course of calculating the balance of energy between the supposed oscillators and their incoming and outgoing radiation, Planck found he needed to assume the existence of quanta, or certain small divisions of energy, rather than a continuous range of possible energies. He defined a quantum of energy as the frequency of the oscillation multiplied by a tiny number that became known as Planck's constant. He then went on to use those assumptions to solve the black body problem; his mathematical solution predicted the black body radiation spectrum perfectly. Planck himself never advanced a meaningful interpretation of his quanta, and there the matter rested until 1905, when Einstein, building on Planck's work, published his theory on the phenomenon known as the photoelectric effect *(above)*. Given Planck's calculation, Einstein showed that the charged particles—by then assumed to be electrons—would have to absorb and emit energy in finite quanta that were proportional to the frequency of the light or radiation. By 1930, quantum principles would form the foundation of a new physics.

The wave theory of light seemed at odds with the idea of energy in packet form, but there was good evidence for the traditional view. Light could be made to produce interference patterns, like the intersecting sets of concentric ripples on the surface of a pond *(page 73)*. Moreover, the wave theory had recently been confirmed by the discovery and application of radio waves, an invisible form of electromagnetism. Planck himself was reluctant to accept the possibility that radiant energy might come in bundles. Gradually, however, he and other theorists came to agree that the theory of black body radiation rested on some sort of discontinuity—whether in the relationship between electrons and radiation or in light itself.

Ultimately, these developments converged to produce a new model of the atom. But the process was piecemeal and not always direct. One of the guiding spirits of the expedition into atomic structure was Ernest Rutherford, whose long research career yielded one significant discovery after another. A New Zealander born in 1871, Rutherford was one of a dozen children. His mother, a schoolteacher who had emigrated from England, taught him to revere education. His father, a Scot who made his living cutting railroad ties, building bridges, and running a flax mill and small farm, bequeathed a technical handiness that would later serve him well: In the laboratory, where funds were limited and ingenuity was required, Rutherford demonstrated an unerring ability first to choose problems that would lead to important answers and then to jury-rig affordable experimental devices to chase those answers down.

In 1895 Rutherford journeyed to England to study at Cambridge. There he worked with Joseph John Thomson, known as "J.J.," head of the Cavendish Laboratory. Within a few years, Thomson would discover the electron and, in consequence, set forth a model of the atom that likened it to a plum pudding—a diffuse, positively charged sphere studded with solid, negatively charged electrons. Thomson's model would hold sway for nearly a decade: Rutherford was destined to be the agent of its demise.

Upon leaving Cambridge, Rutherford took a position at McGill University in Canada, where he would spend nine years and win renown for his studies of radioactivity. Among the subjects he investigated were alpha particles, tiny bodies (which he later showed to be helium ions) that are emitted by some radioactive elements during their decay. By 1907 he had made such a name for himself that Arthur Schuster, head of the physics department at the University of Manchester in England, resigned his post so that Rutherford could be enticed to take it up. The garrulous and enthusiastic Rutherford presided there for a dozen years, building an international center whose reputation rivaled that of Thomson's Cavendish Laboratory.

An admonitory sign, hung by staff scientists at the renowned Cavendish Laboratory in Cambridge, England, reminds the exuberant director Ernest Rutherford to modulate his booming voice. Rutherford, shown on the right chatting with a researcher, could disturb sensitive equipment there simply by speaking in his normal tones.

(Rutherford himself would eventually head Cavendish from 1919 to 1937.)

Two years after arriving at Manchester, Rutherford noticed that in his alpha particle experiments the particles were somehow being deflected slightly when passing through thin gold foils. He suggested that Ernest Marsden, a promising undergraduate student, look into this anomalous scattering behavior. Marsden found that an occasional alpha particle would ricochet off the foil instead of penetrating it. "It was," as Rutherford put it later, "almost as incredible as if you fired a fifteen-inch shell at a piece of tissue paper and it came back and hit you." In 1911 the New Zealander advanced an explanation: The reason most alpha particles zip through the gold foil is that atoms are mostly empty space. In fact, Rutherford concluded, atoms were less like plum puddings than like solar systems in miniature. The center, or nucleus, was a tiny "sun" that contained most of the mass in the system—99.98 percent of it—and carried a large charge; the electrons, negatively charged, orbited like planets at a distance of about 10,000 nuclear diameters. The reason a few alpha particles bounce back is that they are deflected by the dense, highly charged nuclei. Rutherford was uncertain at first whether the charge on the nucleus was positive or negative. Later he and his coworkers demonstrated that the nucleus is made up of two components: positively charged protons, and particles with no charge, called neutrons.

The Rutherford atom was revolutionary. It was also in conflict with some basic principles of physics. According to the theory of electromagnetism painstakingly developed by Scottish physicist James Clerk Maxwell and others in the 1850s and 1860s, accelerated electrons, traveling on curved paths, lose energy by radiation. By this rule, Rutherford's circling electrons could not remain in orbit indefinitely but should rapidly exhaust themselves and spiral into the nucleus, destroying the atom.

A DANE'S RESOLUTION

This quandary was relatively short-lived as such things go. Within a couple of years the beginning of a resolution was put forth by Niels Bohr, a shy young Danish theoretician who came to Manchester for a few months in 1912. Bohr had just spent a half year at Cavendish on a fellowship, hoping to get his exhaustive thesis on electrons translated from Danish into English. When Thomson showed little interest in Bohr's paper, the Dane headed off to Manchester to study radioactivity under Rutherford, with the aim of teaching the subject when he returned to Denmark.

Thus began an unlikely, lifelong friendship. The two men were an odd match. Rutherford's booming voice reverberated in the labs where he worked. Bohr never spoke much above a whisper. Nonetheless, talking was essential to his being. He not only spoke three languages, but, like Shakespeare's tortured Dane, he wrestled with words, reversing and correcting himself, wriggling into and out of paradoxes, repeating himself, searching for exactly the right phrases. If speaking was difficult, writing was excruciating: He wrote drafts even of postcards and would revise papers a half dozen times, driving collaborators to distraction. The complexity of his own intellectual life may have enhanced Bohr's receptivity to the atom that Rutherford had crafted—an atom that made sense experimentally but could not exist under the laws of classical physics. In a bold move, the young physicist circumvented this problem by simply declaring that motions within atoms are governed by other laws. In particular, he contended that electrons do not radiate energy when they are in certain "stationary states."

In 1913 Bohr revealed his vision of the atom in three papers that ran in the British *Philosophical Magazine,* using Planck's constant and the spectral emissions of the hydrogen atom as brush and canvas. He saw that Planck's constant held implications for atomic structure. If energy could be released only in fixed amounts, then it was reasonable to assume that the energy of electrons was subject to limitations. Instead of whirling willy-nilly about the atomic nucleus, electrons must occupy specific orbits, determined by their energy levels. The size and number of orbits would vary for each type of atom, and electrons could absorb or release energy only by jumping from one orbit to another. An electron moving from a higher orbit to a lower one would emit a characteristic quantum of energy; an electron in a lower orbit could absorb only those quanta of energy that would kick it precisely into a higher one. This explained why, for example, atoms of hydrogen give off distinctive wavelengths of light, which show up in the hydrogen spectrum as a fixed distribution of bright lines, known as the Balmer series: The atoms emit energy only in certain precisely calibrated amounts.

Most established scientists were taken aback by Bohr's atom and its im-

Jouncing across a Copenhagen field, Danish physicist Niels Bohr of the Institute for Theoretical Physics test-drives George Gamow's motorcycle, with Mrs. Bohr as passenger. By the early 1930s, the institute was a magnet for young theorists like Gamow who were eager to explore Bohr's notion that electrons travel in orbits determined by their energy levels.

plications for classical theory. But Rutherford sang his praises, calling him "the most intelligent chap I have ever met," and several ambitious young physicists followed his lead. In England and Germany, as well as in the Netherlands, Denmark, and Sweden, a new generation of researchers began developing powerful evidence in support of Bohr's ideas. When Bohr returned to Copenhagen at the end of World War I, he commanded enough respect to garner funds for a new center called the Institute for Theoretical Physics. Along with the German universities of Munich and Göttingen, the institute emerged as a leader in atomic theory. Soon physicists engaged in eager debate were shuttling from one to the other of the three intellectual hubs.

Despite its growing acceptance, Bohr's atomic theory had severe short-comings. While calculations based on the theory held quite nicely for the hydrogen atom, they failed to account for the spectra of other elements. The behavior of atoms with more than one electron was clearly too complicated to be described by Bohr's simple model. In 1925 three researchers at Göt-tingen—a twenty-three-year-old student named Werner Heisenberg; his academic supervisor, Max Born; and another student, Pascual Jordan—set out to construct a mathematical foundation for the study of atoms. Heisenberg, who came from a prominent academic family, was a romantic in spirit, well-versed in the Greek classics but utterly lacking experimental knowledge in physics. His theoretical grasp was unimpaired, however. With a certain arrogance, he surveyed Bohr's quantum theory and dismissed the notion of electrons jumping back and forth between so-called orbits. Did the planets do that? Of course not. Then orbit was not the right concept. Inexact language was getting in the way of understanding, Heisenberg thought.

Returning to the hard data of spectral lines, he laid out the evidence in a form known as a matrix. Like a mileage chart of distances between cities, the matrix listed possible electron "states" (the term he preferred to "orbits") across columns and rows. Each entry in the matrix consisted of a symbol representing the intensity and the frequency of the spectral line that an electron would emit or absorb in jumping from, say, state 1 to state 2, or state 10 to state 9. Using an algebraic technique that allowed him to multiply matrices of different attributes, such as energy or momentum, and with mathematical assistance from Born and Jordan, Heisenberg developed a way to calculate the spectral properties of atoms. He could actually predict the characteristics of the spectral lines that would be emitted by electrons of any atom as they jumped from one energy state to another. This had never been done before. Remarkably, it would soon be done again—in a theory whose approach was completely different from Heisenberg's.

The new hypothesis was proposed in January 1926 by Erwin Schrödinger, an Austrian who taught in Zurich, Switzerland. Two years before, a French theoretician, Louis de Broglie, had suggested that electrons within the atom could be described not only as particles, as in Bohr's theory, but also as waves. Schrödinger elaborated on this concept, arguing that electrons were not objects orbiting the nucleus at all; rather, they somehow smeared out into waves. Banishing the electron-as-particle, Schrödinger contended that changes in energy output were caused not by hopping electrons, as Bohr had said, but by shifts from one kind of wave pattern and frequency to another. Many physicists embraced his theory of wave mechanics, finding it considerably easier to visualize. It also involved mathematics far less complicated than those required for Heisenberg's matrix approach.

A rousing battle ensued, with Bohr and Heisenberg on one side and Schrödinger—who had the support of both Planck and Einstein—on the other. The debate ended more than a year later, settled by the application of a kind of cosmic compromise that would become well known as the Heisenberg uncertainty principle *(pages 70-71)*, published in May 1927. The reason it was not possible to say with certainty whether subatomic particles were waves or particles was because . . . it was simply not possible.

This was no mere tautology. Rather, it was a statement of our fundamental inability to know the universe completely. The uncertainty principle arose from Heisenberg's realization that the very act of attempting by experiment to establish the position of an electron relative to, say, the nucleus would change the velocity of that electron in a way that could not be predicted. An electron could only be described in terms of probabilities: Pinpointing its location would render its precise speed or direction unknowable; determining its speed and direction would preclude pinpointing its location. Bohr elaborated on the paradoxical theme. The critical implication of the uncertainty principle was that—depending on how the experiment was set up—electrons at any given instant would act like waves or like particles, but not both.

French aristocrat Louis de Broglie won the 1929 Nobel physics prize for a doctoral thesis elucidating the wavelike properties of orbiting electrons. The work helped solve a longstanding paradox by showing that electrons can be described in terms of both particles and waves, depending on the circumstances.

ACCOUNTING FOR SPIN

The unpredictable behavior of particles in the quantum world seemed to bear little relation to the behavior of bodies at the larger scale of relativity theory. Both theories were still evolving, and efforts to combine them were only partly successful. For example, none of the attempts at synthesis could adequately account for a recently discovered property of electrons called spin, proposed to resolve observed anomalies in the positions and numbers of lines in atomic spectra. At the time, physicists thought that a rapidly rotating electron created a magnetic field, which would account for otherwise mysterious shifts. But to produce these magnetic effects, an electron with the dimensions assigned to it in classical theory would have to rotate so fast that points on its equator would exceed the speed of light—something that relativity theory said was impossible.

Paul Dirac, an English theorist who worked at Cambridge, rose to the challenge. The son of a Swiss immigrant who taught languages at a Bristol

IN THE REALM OF UNCERTAINTY

Perhaps the most fundamental tenet in the notoriously fuzzy study of quantum physics is the so-called Heisenberg uncertainty principle, the concept that the very act of observing changes the thing being observed. In 1927 German physicist Werner Heisenberg realized that the rules of probability governing subatomic particles are borne of the paradox—depicted in the thought experiments shown here—that two related properties of a particle cannot be precisely measured at the same time. For example, an observer can determine either a particle's precise position in space or its precise momentum (the product of velocity and mass) but never both simultaneously. Any attempt to measure both results in imprecision.

As it happens, this rule applies at all scales of matter and motion. The photons from a police radar intended to clock a speeding car, for instance, actually affect the vehicle's speed, but because photons are so small compared to the car, the disturbance is imperceptible. In the subatomic world, however, photons are the tools for studying the properties of particles nearer their own size. The very act of measuring thus interferes violently with the thing being measured.

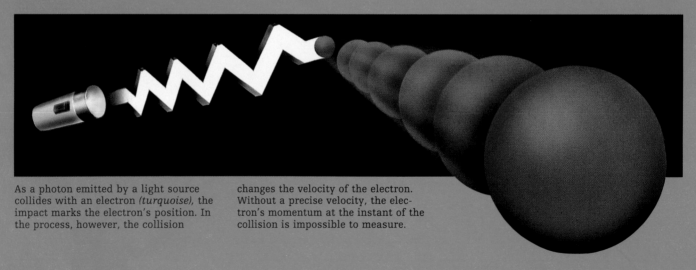

As a photon emitted by a light source collides with an electron *(turquoise)*, the impact marks the electron's position. In the process, however, the collision changes the velocity of the electron. Without a precise velocity, the electron's momentum at the instant of the collision is impossible to measure.

grammar school, Dirac was fluent in French and English, but taciturn in both. His true tongue was mathematics, and in 1928 he succeeded in incorporating relativity into a mathematical description of the mechanics of a hydrogen atom. His solution, called the Dirac equation of the electron, not only yielded a perfect explanation of the spectral lines but, in an unlooked-for development, also described electrons in a way that resolved the spin dilemma.

The simple elegance of Dirac's mathematics won his proposal quick acceptance. Meanwhile, other physicists were forging ahead into the next unexplored territory: the nucleus, then considered to be made up exclusively of neutrons and protons. In the mid-1930s Hideki Yukawa, a graduate student and lecturer in physics at Osaka University in Japan, theorized the existence of the so-called strong force. Far stronger than electromagnetism, this force holds protons and neutrons together and is effective only over the extremely short range of the nucleus itself. Yukawa went on to maintain that the strong force is conveyed between protons and neutrons by the exchange of unknown

In this classic thought experiment, a box filled with radiant energy hangs from a scale calibrated along its vertical support post. Inside is a clock that controls a shutter. The box is weighed, and then, at a certain time, the clock opens the shutter, allowing a photon of light to escape *(right)*. When the box is weighed again, the difference in weight indicates the photon's exact mass, which in turn yields a measure of its energy.

However, the procedures needed to measure the photon's energy prevent an observer from knowing exactly when the particle was released. This is because the rate at which a clock runs depends on its position in a gravity field *(pages 50-51)*. In the instant the photon is released, the box weighs less and moves up the scale, away from Earth's gravity—changing the clock's rate by an indeterminate amount.

particles heavier than electrons. The search for these bridgelike carriers of the strong force began immediately. Within a few years a candidate was found and designated the meson. This turned out to be a case of mistaken identity, however. The true carriers of the strong force, given the name gluons, have still not been detected.

A QUANTUM KEY

By the end of the 1930s, theorists had assembled a respectable mathematical arsenal for analyzing the universe at the subatomic scale and, separately, at the scale of planets, stars, and galaxies. They now understood the atom to consist of energetic electrons surrounding a nucleus made up in turn of protons and neutrons. But they also realized, with varying degrees of comfort, that the behavior of this tiny system was subject to quantum principles exceedingly different from those operating in the world at large. The question that had tantalized Georges Lemaître—how to get, mathematically, from a very small, dense source to an expanding universe of galaxies—had yet to be answered. To some, the key to describing the processes governing the early universe seemed to lie in the strange subatomic world.

The specific conditions of the infant cosmos, whose reality was still a subject of intense scientific debate, were first postulated by the Russian-born scientist George Gamow in the late 1940s. In 1923, Gamow studied with Friedmann at the University of Petrograd (now Leningrad). By this time Friedmann had solved Einstein's field equations and theorized variations on an expanding universe. He not only taught his student cosmological principles but alerted him to the strides that were being made in quantum physics.

With trancelike absorption, Niels Bohr *(far left)* and Albert Einstein parley at a physics conference in Brussels, around 1930. Until Einstein's death in 1955, the two scientists contended over the merits of Werner Heisenberg's uncertainty principle. Bohr endorsed Heisenberg's assertion that some things in the universe are beyond knowing; Einstein found that view insupportable.

THE RIDDLE AT THE HEART OF PHYSICS

The challenge of trying to understand the nature of matter and energy has taxed the wits of scientists since the days of Aristotle. Many have looked to the study of light for answers—but often the answers only raised more questions. In the seventeenth century, for example, Isaac Newton experimented with light-refracting prisms and propounded the notion that light was made up of a stream of particles. This view was widely accepted until the early 1800s, when Newton's countryman Thomas Young found strong reasons to reject the light-as-particle theory.

Using a device similar to the ones that are illustrated below, Young shone light through narrow slits in a board toward a detecting screen set up behind the board. The result—a series of bright and dark strips—convinced him that light moves in waves. Just as openings in an ocean breakwater generate overlapping wave sets, the slits in Young's experiment seemed to cause light to produce a similar interference pattern *(below, right)*. The bright bands on the screen represented places where the waves reinforced each oth-

er; the dark bands, where the waves canceled out.

Young's interpretation supplanted particle theory for a while but could not explain phenomena such as the photoelectric effect *(page 64)*. As it happens, neither wave nor particle theory alone can completely describe the behavior of light. It takes the judicious application of both models to account for electromagnetic phenomena—a paradox that has been dubbed the "central mystery" of physics.

Scientists who perform Young's double-slit experiment today with more elaborate equipment can actually demonstrate light's dual nature. As illustrated below, particles such as bullets would produce one kind of pattern when passing through two slits, waves another. If light is particulate, the pattern it creates should resemble that generated by bullets. And in fact, if first one slit and then the other is closed, the resulting pattern follows the rule of particle behavior: two bands aligned with the slits. But as soon as both slits are open, the pattern switches to the series of stripes resulting from wavelike interference. This holds true even if photons are fired at the screen so slowly that they reach the screen one at a time. It is as if each particle were somehow passing through both slits. As yet, not even Nobel physicists can explain this peculiar manifestation of quantum weirdness.

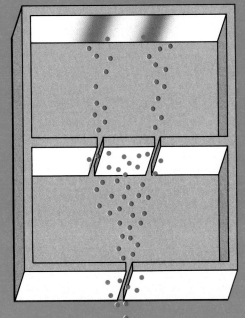

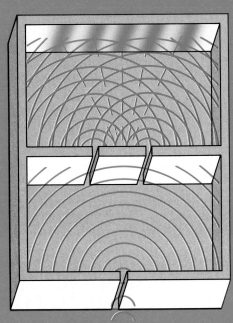

In the experiment shown at far left, bullets from a machine gun pass through two slits in the barrier and pepper the rear screen. In time, two bands of hits emerge. The riddle of light is that this pattern will be created if first one slit and then the other is closed. But if both are open, light will generate the wave interference pattern at left.

At left is the interference pattern of narrow bands typically produced by waves. Surging through the narrow openings, waves split into new sets of concentric ripples that cross and recross before hitting the rear screen. The dotted lines mark intersecting crests, reinforcements that show up strongly on the detecting screen.

After receiving his Ph.D. in 1928, Gamow went to Göttingen in Germany. The focus of his work there was alpha particles, Rutherford's instrument for probing atomic structure. Because an alpha particle—a tiny system made up of two protons and two neutrons—is normally confined to the atomic nucleus, physicists had puzzled over its ability to break out of the nucleus during radioactive decay without the system itself breaking apart. Gamow achieved a significant research coup at Göttingen. Using Schrödinger's wave theories, he showed that particles in certain excited energy states could, in effect, tunnel their way out of the nucleus. The work attracted the attention of many physicists, including Niels Bohr and Ernest Rutherford. In 1929, while Gamow was at Cavendish on a Rockefeller fellowship, Rutherford asked him to investigate whether, since particles could get out of the nucleus, they might somehow be induced to break into—and thus split—it as well.

On Rutherford's suggestion, Gamow considered the effects of bombarding atomic nuclei with protons artificially accelerated to enormous speeds in a high-powered magnetic field. In time his calculations showed that tunneling into the nucleus was possible and that accelerated protons would stick to the nuclei, forming new nuclei. Moreover, he determined that the acceleration need not be as high as that required under the laws of classical physics, a finding that encouraged researchers at Cavendish to build a particle accelerator themselves. In 1932 that instrument succeeded in disintegrating lithium and other light nuclei with protons, a major breakthrough that opened up the study of nuclear physics.

OUT OF RUSSIA

In 1931, after a stint at Bohr's institute in Copenhagen, Gamow returned to the Soviet Union to renew his visa and found himself effectively grounded for a couple of years. Chafing under the strictures of the increasingly repressive Stalinist regime, he determined to seek his intellectual fortunes elsewhere. In 1933 the frustrated scientist managed to persuade the authorities to let him attend a conference in Brussels with his wife. He did not return. After several months of itinerant teaching in both Europe and the United States, Gamow settled into the faculty at the George Washington University in Washington, D.C. In that haven, over the next two decades, he extended his contemplations from bombarded atoms to the very origins of the universe.

Now the Russian theorist concentrated on the primeval atom conceived by the Belgian theorist-priest. Gamow called Lemaître's atom the cosmic singularity. If matter and energy were interchangeable, as Einstein had shown with his famous equation $E = mc^2$, the enormous densities and high temperatures at the beginning of time and space—what Gamow dubbed the Big Squeeze—would have smeared the distinction between them. The chief component of that hot matter-energy stew, Gamow suggested, would have been radiant energy. Then, as the universe began to expand and cool, the first matter to emerge would have been in the form of elementary particles: protons, neutrons, and electrons. One of Gamow's later collaborators, Ralph

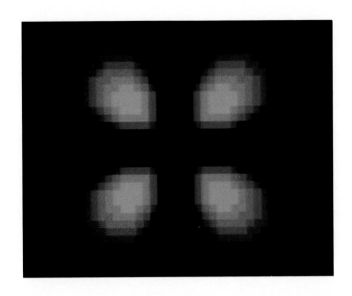

A computer-simulated slice through a hydrogen atom in an excited state portrays the probable haunts of the atom's single orbiting electron. (The brightest regions signify its most likely positions.) According to the laws of quantum mechanics, a look at the same atom in any of an infinite number of other energy states would show a changed pattern of probabilities.

Alpher, gave the name "ylem," a term taken from Aristotle, to this primordial matter. From his Cavendish studies Gamow had learned that neutrons, although they are stable within atomic nuclei, rapidly disintegrate when on their own. A solitary neutron has an average lifespan of only thirteen minutes before it breaks into a proton and an electron. In the hot density of the infant universe, neutron decay would have been balanced by the production of new neutrons from the collision of protons and electrons.

But as the ylem cooled and grew less dense, fewer and fewer collisions resulted in fewer and fewer new neutrons. At the same time, there was a reduction in high-energy radiation, allowing existing neutrons to begin to combine with protons, forming atomic nuclei as they did in the particle accelerator at Cavendish. Lone protons attracted electrons to make hydrogen atoms, and the heavier nuclei also gathered their larger complements of electrons. The Big Squeeze could be the crucible for all the elements now observed in the universe. By the mid-1940s Gamow had teamed with Alpher, then a George Washington graduate student, to carry out the difficult computations that revealed how this might be done.

Gamow and Alpher produced a monumental paper but gave it a rather whimsical debut in a letter that, quite by coincidence, appeared on April Fools' Day, 1948, in the journal *Physical Review*. Gamow, a dedicated practical joker, added the signature of renowned atomic physicist Hans Bethe, even though Bethe had had nothing to do with the project. (Luckily, Bethe took the joke well.) The resulting list of authors—Alpher, Bethe, and Gamow—was a pun on the first three letters of the Greek alphabet, alpha, beta, and gamma. (Gamow and Alpher were later joined by Robert Herman, a colleague of Alpher's, forming a collaboration that persisted even after Gamow moved to the University of Colorado in 1956.)

Gamow's *Physical Review* gag was appropriate, since the journal letter presented a hypothesis for the creation of the first elements in the periodic table. It outlined the process by which neutrons and protons collided, agglomerated, and were reconfigured to make, first, heavy hydrogen (one neutron plus one proton) and then tritium (two neutrons plus one proton). In the next step, when one of the neutrons in the tritium nucleus decayed into a proton and an electron escaped, the tritium nucleus was transformed into an isotope of helium called helium-3. Adding one more neutron to helium-3 produced the isotope helium-4.

Here was the very stuff of alchemy: In the ylem, one substance was transmuted into another. Elaborations by other researchers have since revealed that the events described by Gamow and Alpher transpired within the first few minutes of cosmic expansion. The processes they outlined produced the proper abundances of hydrogen and helium, which constitute all but about one percent of the matter in the universe.

Gamow also accounted for the observed distribution of mass throughout space. For the first million years, he said, radiation ruled the universe. By then, enough matter had formed for it to become predominant, and about 240,000,000 years later great billows of gas began to coalesce. The mass of these gas clouds, he figured, was roughly equivalent to that of today's galaxies.

In the course of his career, Gamow's flair for the light touch brought him considerable success in the book world and, in the process, led his esoteric theory out of the ivory tower and into the public consciousness. *The Creation of the Universe*, published in 1952, became a bestseller. But his

George Gamow, shown here with his cat Spin, championed the Big Bang theory for nearly three decades. In the 1940s the Russian-born physicist and his colleague Ralph Alpher described how elements may have formed in the first moments of cosmic time.

vision did not go unchallenged. In exploring the synthesis of elements, Gamow and Alpher had encountered a major stumbling block. Try as they might, they could not explain the manufacture of elements heavier than helium-4, an isotope so stable that it all but refuses to accept or give up particles and thus engender weightier atoms. In any case, by the time helium-4 nuclei would have been created—a few minutes into expansion—the cosmic soup of particles would have thinned so much that collisions would not have occurred often enough to forge the heavier elements.

A further deficiency of Gamow's hypothesis became evident when the observed expansion of the universe was used to gauge the amount of time that had elapsed since the moment of creation. If the universe could be thrown into reverse at the velocity derived from observation—equivalent to running a movie backward—the cosmos would converge in a singularity at a moment about 1.8 billion years in the past. Yet geologists, using lead isotopes to date rocks, had marked the oldest formations on Earth at about four billion years, more than twice the alleged age of the whole universe.

ENTER THE STEADY STATE

This contradiction opened the door for a competing hypothesis. Called the Steady State model, it was championed by Fred Hoyle, a British astronomer, and supported by two Austrian-born colleagues, Thomas Gold and Hermann Bondi. The trio proposed that the universe had no definite beginning but was infinite, both temporally and spatially. During a BBC radio program, Hoyle introduced the Steady State theory and coined the catchy phrase "Big Bang" to describe his rivals' model, thereby christening each side of a dispute that would continue for two decades.

Gold and Bondi had been interned together in Britain as enemy aliens during World War II. Bondi, an outstanding mathematician, began tutoring

Shunning the notion of a cataclysmic cosmic birth, British mathematician Fred Hoyle helped formulate the Steady State theory, the main rival to George Gamow's Big Bang model. Hoyle held that matter came into being continuously over time in a steadily expanding universe.

his companion. Without this assistance, Gold later said, "I would have been a complete dilettante at the business." Late in the war the British government decided to take advantage of their skills and assigned the pair to help the admiralty develop a radar system. In a shared cottage near a Canadian Air Force base north of London, they spent many off-hours in lengthy discussions about the nature of the cosmos, trying to ignore the explosions thundering nearby as returning bombers dropped their unused cargo on a dumping field.

Hoyle, who had recently completed graduate work in mathematics and theoretical physics at Cambridge (supervised for a time by Paul Dirac), joined Bondi and Gold on the radar project and was soon included in the cosmological bull sessions. Hoyle was puzzled—obsessed even—by findings on the red shift of light from distant galaxies *(pages 22-23)*. Most scientists took the red shift as clear evidence that the universe was expanding, but Hoyle was not so sure. "Well, what could that observation mean?" he would ask his companions. "Find out what it could mean!"

Although he was himself a brilliant mathematician and lectured in the subject at Cambridge for years, Hoyle often acted as the idea man, delegating detailed calculations to others. In this case Bondi did most of the work. Gold later recalled his friend sitting cross-legged on the floor while Hoyle sat behind him in an armchair, kicking him "every five minutes to make him scribble faster, just as you might whip a horse." Bondi, not always certain where Hoyle was driving him, once looked up and asked, "Now at this point do I multiply or divide by 10^{46}?"

After the war the sessions shifted to Cambridge, where one evening in 1946 or 1947—neither Bondi nor Gold can remember which—Gold began musing about whether a big bang was really necessary. "Do we really know the universe must have come from a beginning?" he asked. "After all, we had to create matter then, at the beginning. Why not create it a little at a time?"

Hoyle and Bondi dismissed these maunderings. Says Bondi, "We thought Tommy's first idea was crazy and were sure that we could shoot it down in next to no time." What about the law of conservation of energy? they inquired. Certainly, creating matter here and there a little at a time was a contradiction of the law that said that the total energy in a closed system remains the same. "Well, look," replied Gold, "if you make it in one big bang to begin with, you are not conserving energy either." Having matter appear gradually seemed preferable, in part because it sidestepped (or at least muted) the question of a Creator. As Bondi and Hoyle examined the issue further, they found themselves increasingly convinced.

However, as time went by, an intellectual chasm opened between Bondi and

Gold on one hand and Hoyle on the other. Hoyle attacked the problem from a mathematical angle, while Bondi and Gold's approach was more philosophical. Working separately on his own version of the theory, Hoyle asked if he could list Gold as coauthor, but Gold declined and spurred Bondi to finish a paper the two of them had begun on the subject. Hoyle first sent his piece to a physics journal, but the editors suggested that he try the *Monthly Notices* of the Royal Astronomical Society instead. By then Bondi and Gold had completed their paper, which appeared in the *Monthly Notices* on July 13, 1948. Hoyle's followed on August 3.

Bondi and Gold argued for the Steady State theory on the basis of what they called the perfect cosmological principle. The original version of the cosmological principle, fundamental to the Big Bang theory, holds that for any observer at any point in space the universe appears the same. The perfect version expands the parameters to include time, which means that the universe should appear the same at any moment—past, present, or future. The Big Bang universe clearly contravened that principle. Therefore, in the opinion of Bondi and Gold, it had to be discarded. And therefore, the state of the universe must be steady, maintained by the constant creation of matter in the form of hydrogen atoms.

Hoyle's argument derived mathematically from a modification of general relativity. His equations produced an expanding universe with a constant density. Unlike Bondi and Gold, Hoyle did not specify the form of matter needed to maintain density within an increasing volume; in his version, undifferentiated matter-energy is created at a rate linked in the equations to the rate of expansion. To those who might ask why, if matter is being created, no one has seen it happen, the Steady Statists countered that a minuscule rate of creation would suffice. The appearance of one or two new atoms each year for every 35 million cubic feet of space (about the size of the Empire State Building) would do the trick. In any event, Hoyle declared, the admittedly startling idea of continuous creation was more palatable than the assumption that all matter in the universe was created at one time in the remote past, in some manner that could not be described scientifically.

Throughout the 1950s the proponents of the Steady State model, led by Hoyle, and those of the Big Bang, urged on by Gamow, wrangled tirelessly with each other. At the outset, as one physicist said later, neither camp could "bring much to bear in the way of convincing evidence for one theory or the other." Rather, "they spent a good deal of time insulting one another, something at which both Hoyle and Gamow were expert."

But in that decade, the Big Bang achieved a major victory. Reexamination of astronomical evidence indicated that the distances to other galaxies had been significantly underestimated. Within a few years those distances—and thus the age of the universe—were revised upward by a factor of five. Initially thought to be scarcely two billion years old, the universe was now estimated to be ten to twenty billion years old. One of the knottiest inconsistencies of the Big Bang theory was removed.

ACCOUNTING FOR THE PERIODIC TABLE

Undaunted, Hoyle attempted to capitalize on a different failure in Gamow's theory: The Big Bang could not explain the production of heavy elements. Since about 1946 Hoyle had been contemplating the possibility that elements were formed by nuclear reactions, or nucleosynthesis, within stars, and he had published a paper elaborating the hypothesis. When he began working on the Steady State theory, he grew increasingly convinced that the spontaneous creation of matter in some elementary form, together with the engendering of other matter in stellar crucibles, would account for all elements in the periodic table. There would be no reason for the Big Bang at all.

In late 1954 Hoyle joined a group that was investigating this question. Geoffrey Burbidge, at Cambridge's Cavendish Laboratory, had been working for some time on the relative abundance of elements on the surface of stars. One partner in the effort was his wife, Margaret (who later was the first woman appointed director of the Royal Greenwich Observatory). Another was William Fowler, a specialist in low-energy nuclear physics who was visiting Cambridge from the California Institute of Technology.

At Cambridge and later at Caltech, the group carried out an exhaustive study of element building in stellar cores. Their monumental account, "Synthesis of the Elements in Stars," ran to 100 pages in the October 1957 issue of *Reviews of Modern Physics.* "We have found it possible," they wrote, "to explain, in a general way, the abundances of practically all the isotopes of the elements from hydrogen through uranium." As stars used up their nuclear fuel, they transmuted one element into the next: The "ashes" of hydrogen were helium; the ashes of helium, carbon and oxygen; and so on up the periodic table. In the catastrophic explosions of some giant stars—events known as supernovae—many of the heaviest elements were made, including platinum, gold, and uranium. B^2FH, as colleagues dubbed the seminal paper (for Burbidge, Burbidge, Fowler, Hoyle), not only accounted for the synthesis of all the elements beyond hydrogen, it predicted their formation in just the same proportions as they occurred in the universe.

Although the creation of hydrogen itself was still an open question, the work was an important scientific achievement, and the whole team was proud and exhausted at the end of it. As the paper was being prepared for publication, the group embarked on a short expedition to the Scripps Institute in La Jolla, a two-hour drive from Caltech. Scripps was trying to recruit the Burbidges, but another purpose of the trip was to visit Hoyle's old nemesis, George Gamow, who was doing a stint in La Jolla as a consultant to the defense firm General Dynamics. As Hoyle later recalled, Gamow had "blown every penny" of his yet-to-be-earned consulting fee, most of it on a huge white Cadillac convertible. Gamow and Hoyle took a spin around La Jolla, bickering still. During the mobile debate, Hoyle thought to score further points off Gamow by raising the subject of cosmic background radiation. This, in his judgment, was another flaw in the Big Bang hypothesis. But events would prove him wrong.

In 1965, using this horn radio antenna in Holmdel, New Jersey, Bell Telephone Laboratories astronomers Arno Penzias and Robert Wilson unwittingly detected the microwave radiation that represents the faint echo of the Big Bang. They thought they had picked up random noise, but theorist Robert Dicke at nearby Princeton University recognized the signal for what it was.

Cosmic background radiation had been predicted by Alpher and Herman several years earlier in a 1948 letter to the journal *Nature.* If, as Gamow estimated, the temperature of the universe had been a billion degrees three minutes after the Big Bang, then—like a pot taken off the burner and set aside to cool—the cosmos should still show signs of its early superhot phase. Because of the continued expansion of the universe over billions of years, this high-energy, short-wavelength radiation, which occupied all of space-time, would gradually stretch, or red-shift, becoming low-energy, long-wavelength radiation. Alpher and Herman calculated that space should now be awash in a sea of electromagnetic energy that, in black body terms, has a temperature of about five degrees above absolute zero, or five degrees Kelvin.

Here was that rarity in cosmology—a testable hypothesis. Radio astronomy was a fledgling science in 1948, but it would have been possible even then to turn an antenna on the heavens and take a reading. Radiation at five degrees Kelvin would register in different amounts at different wavelengths in the microwave range, from three centimeters to one-hundredth of a centimeter.

But Alpher and Herman—though they privately discussed the feasibility of such an experiment—never publicly suggested a follow-up. No one else bothered to look for the signals either: Most observational astronomers were unaware of the prediction because they never saw the article. "It was one of those circular things," said one physicist, commenting on the missed chance. The field of cosmology "was not taken seriously because of a lack of crucial data, and there was a lack of crucial data because it was not taken seriously."

Certainly Hoyle and Gamow took cosmology seriously, but they failed to notice the data even when it stared them in the face. During their Cadillac tour of La Jolla, Hoyle remembered, he told Gamow "that it was impossible for the universe to have a microwave background with a temperature as high as he was claiming." Hoyle based his argument on some 1940 studies showing that the vast clouds of material floating in the reaches of interstellar space had a temperature of about 2.3 degrees Kelvin. If this, the coldest matter known, was only 2.3 Kelvin, Hoyle argued, how could the remnant radiation from the Big Bang, which had been cooling for billions of years, be as warm as Gamow suggested? What had not occurred to Hoyle was how, out in the cosmic equivalent of Antarctica, the clouds had any heat to give off at all. Only much later did he realize that background radiation would warm them in passing and that therefore the temperature of the clouds would be very close to that of the background. Said Hoyle ruefully, "I was evidently not 'discovery prone.'"

Hoyle was not alone. Robert Dicke, a physicist at Princeton University, had actually detected cosmic background radiation in 1946, two years before it was predicted. Working with a special measuring device he had developed, Dicke discovered radiation with a temperature somewhere below twenty degrees Kelvin, the lower limit of his instrument's precision. He reported this finding in a paper that appeared in the *Physical Review* but, having no explanation for the phenomenon, he then put it out of his mind.

Dicke did not recall this observation when he began doing cosmological modeling a couple of decades later in the early 1960s. Although he knew next to nothing about the initial Big Bang theories and predictions, he reached the conclusion on his own that there should be fossil radiation left over from the infancy of the universe. He asked a fellow Princeton researcher, James Peebles, to see what he could produce in terms of hard numbers, and Peebles came back with a figure of ten degrees Kelvin for the ambient radiation. Dicke next asked two other Princeton researchers, Peter Roll and David Wilkinson, to see if they might be able to find the missing radiation. To that end, Roll and Wilkinson built an antenna on the roof of the Princeton geology building.

In the meantime, not far away, a pair of radio astronomers at Bell Telephone Laboratories near Holmdel, New Jersey, had stumbled onto something peculiar. Arno Penzias was a Polish Jew whose family escaped from Germany just before World War II broke out. He studied physics at City College in New York, got his doctorate from Columbia in 1961, and then took a job with Bell Labs. In 1963 he was joined by Robert Wilson, a Texan who had studied at Rice University in Houston and at Caltech in Pasadena. As it

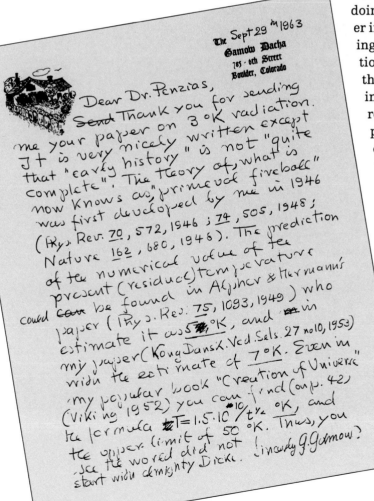

In a note to Arno Penzias (misdated by two years), a miffed George Gamow lists the many articles in which he and others discuss the 1948 prediction of cosmic background radiation. Having treated the prediction even in a best-selling book on cosmology, Gamow was upset that Penzias and his colleague Robert Dicke did not cite his work in their papers claiming discovery of the phenomenon.

happened, Wilson had taken a cosmology course at Caltech from Hoyle, who was there on a visiting lectureship, and the student had become a believer in the Steady State. Like Dicke, both Penzias and Wilson were unaware of Gamow's work. The two scientists had at their disposal a radio antenna that had recently been used in testing some early communications satellites, and they hoped to use the instrument for high-caliber radio astronomy. They therefore undertook to correct certain deficiencies in the antenna, but no matter what they did, they kept getting readings of microwave radiation at about three degrees Kelvin. At this point they thought the problem was caused by pigeons nesting in the horn. But when the birds were evicted, the radiation remained.

By chance Penzias mentioned the trouble to a fellow researcher, Bernard Burke of the Massachusetts Institute of Technology. Burke said that a friend of his had just heard Peebles give a lecture about the cosmological conjecturing at Princeton. Maybe, Burke suggested, the two groups should get together. When they did, Penzias and Wilson reacted with the intellectual equivalent of a shrug: They could not understand why the Princeton boys were so excited about their three-Kelvin excess.

Nonetheless, at the urging of Dicke, Penzias and Wilson wrote up an account published in the July 1965 issue of the *Astrophysical Journal.* The paper attached no significance to the reported finding, instead referring readers to a companion piece from Princeton in the same issue that offered a possible explanation. Neither paper mentioned the key 1948 papers that had predicted the existence of the background radiation, a lack of acknowledgment that understandably upset Gamow and others. Subsequent measurements confirmed the evidence: that the visible universe of galaxies and stars is permeated by radiation at 2.7 degrees Kelvin—the perceptible echo of creation. As far as most scientists were concerned, the Big Bang had won.

When credit for the Big Bang theory was finally apportioned by the Nobel prize committee in 1978, many of the principals had passed from the scene. Lemaître, the man who had started the ball rolling, who had above all else desired to find a cosmological model that fit the real universe, learned about the discovery of the cosmic background radiation shortly before his death in 1966. Gamow died in 1968. Since the prize is awarded only to living scientists, the committee settled on Penzias and Wilson, the two researchers who had found proof for the theory in complete ignorance of Lemaître, Gamow, and the others who had shaped it.

By the time of the award, cosmologists the world over were investigating the Big Bang theory in ever greater detail. The circumstances of the earliest stages of the Big Bang were worked out in 1967 by a team of theorists that included William Fowler and Fred Hoyle. By the 1970s their description was institutionalized as the standard model, the basis for understanding the evolving structure and dynamics of the universe. Lemaître would have appreciated the effort. "Standing on a cooled cinder," he wrote in 1950, "we see the slow fading of suns, and we try to recall the vanished brilliance of the origin of the worlds."

INGREDIENTS FOR A COSMOS

P erhaps the preeminent challenge of science is to work out the recipe for the cosmos, identifying and measuring the stuff from which reality is fashioned. Some ancient Greek thinkers offered an inspired guess, surmising that everything is made of irreducibly small particles that differ only in shape and arrangement. For more than 2,000 years, these particles—called atoms, from the Greek word for "indivisible"—remained speculation. Then nineteenth-century chemists found evidence for real atoms, and the pace of discovery quickened. By the 1920s atoms proved divisible after all, made up of a nucleus of protons and neutrons and surrounded by electrons. In the 1930s machines dubbed atom smashers revealed that at higher energies even more exotic particles appeared. With more powerful equipment the list of so-called fundamental particles grew; by the 1960s physicists had found hundreds, all supposedly elementary.

This profusion was misleading. During the next decade, fresh theories—and new machinery capable of testing them—showed that most subatomic particles are actually varying combinations of a handful of fundamental objects. According to the current view of the subatomic world, illustrated in the pages that follow, these constituents can be grouped into two subclasses called fermions, for Nobel laureate Enrico Fermi, and bosons, named after Satyendra Bose, an Indian physicist. Some fermions occur in isolation; others clump together. All interact by means of the four known forces—gravity, electromagnetism, and the subatomic "strong" and "weak" forces— each conveyed by the bosons. Fermions and bosons inhabit a place where common sense does not always apply. At the scale of these particles, reality becomes inherently fuzzy. The resulting uncertainty can yield strange phenomena, including the apparent creation of something from nothing. Yet the cumulative outcome of these mysterious comings and goings is very familiar: the universe itself.

Packets of Matter

At the most elementary level, matter is thought to consist of fragments, called fermions, so small that more than a trillion would have to be laid side by side to stretch across the width of a human hair. The most fundamental fermions are classified as either leptons or quarks. Leptons are usually solitaries; except under special circumstances, they do not combine with one another or with other particles. Quarks, by contrast, are always embedded in larger, multiquark particles. As illustrated at right, each type of lepton or quark has an antimatter twin that is identical in mass but opposite in other properties, such as electrical charge.

Leptons come in six varieties, three with charge (all negative) and three without. All are low-mass objects even by subatomic standards. Of the charged leptons—electrons, muons, and taus—the heaviest, the tau, has nearly twice the mass of a hydrogen atom. The three types of chargeless leptons, or neutrinos (Italian for "little neutral one"), are such lightweights that most physicists consider them altogether massless. Named for the charged leptons they sometimes accompany, they include the electron-neutrino, the muon-neutrino, and the tau-neutrino. Leptons vary in longevity: The electron, for example, never decays, and the same is believed to be true of all three types of neutrinos. The tau, in contrast, can exist for only a ten-millionth of a microsecond.

The term quarks was plucked by a fanciful physicist from the line "three quarks for Muster Mark" in James Joyce's novel *Finnegans Wake*. (As it happens, quarks do combine in threes.) Usually more massive than leptons, quarks are bound together by the strong nuclear force to make neutrons, protons, and other, rarer forms such as pions and kaons. Indirect evidence suggests that quarks, like leptons, come in six varieties. Physicists, with unabated whimsy, have christened them *up, down, charm, strange, top* (yet to be verified by experiment), and *bottom*. Finally, each of the quarks comes in three versions called colors, making for eighteen quark types in all.

A generic fermion *(below, left)* is paired here with its corresponding antiparticle *(right)*, a bit of antimatter of the same mass but with all other properties reversed.

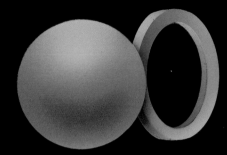

The most familiar charged lepton is the electron *(below, right)*, carrier of electrical current and a key component of the atom. Its antiparticle, the positron *(left)*, has a positive rather than a negative charge.

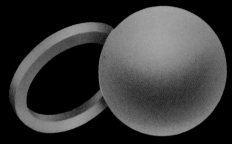

The electron-neutrino, one of three neutrino varieties, is depicted below at left. Both neutrinos and antineutrinos *(right)* are chargeless and probably massless.

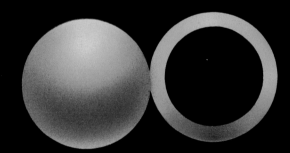

Seen here in isolation, generic quarks *(below, left)* and antiquarks *(right)* are always bound by the strong force to form composite particles with others of their kind.

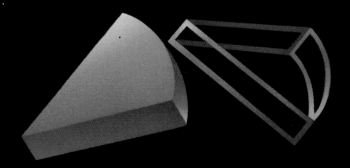

CARRIERS OF FORCE

Bosons are the messenger particles that transmit force from one fermion to another. Each of the four known fundamental forces has its own type of boson, depicted graphically here. Gluons convey the strong nuclear force, which bonds quarks as well as protons and neutrons; eight gluon types differ in abstract properties called color and anticolor. Intermediate vector bosons transfer the weak nuclear force, which changes one type of nuclear particle into another in the process of radioactive decay; these force carriers may be positively charged, negatively charged, or neutral. Gravitons—yet to be detected—are believed to mediate gravity, and photons carry electromagnetism.

Some bosons are stable particles; most photons, for example, date from just after the Big Bang. But force carriers can also be so-called virtual particles, ghostly constructs generated by one fermion and immediately absorbed by another. Through the exchange of virtual bosons, quarks or leptons are able to interact, repelling, attracting, annihilating, or otherwise affecting one another.

The generic gluon at right represents the eight types that bind quarks together. Bearers of the strong force, gluons also bind neutrons and protons to form atomic nuclei.

Intermediate vector bosons come in three distinct types. All of them carry the weak force, which acts on certain kinds of particles that are susceptible to decay.

Photons communicate electromagnetic force between charged fermions such as electrons, allowing them to attract or repel, depending on their respective charges.

Gravitons are thought to convey the tug of gravity between all particles. Although essential to theories that treat the four forces as equivalent, gravitons have yet to be detected experimentally.

Quarks and Gluons in Action

Fermions and bosons can combine in many ways. One of the more fundamental involves the exchange of gluons between quarks in the process of combining the quarks into larger particles such as protons or neutrons. Several natural laws govern this interplay. Most pertain to such abstract quark properties as charge, spin, and color *(table, opposite);* the end result—whether proton or neutron—is also subject to these laws. A typical rule is that color must be "conserved," remaining constant in total value throughout any quark-gluon exchange. Color in this context has nothing to do with the hues of an artist's palette; it is simply a useful term for distinguishing among quarks and gluons on the basis of the ways they influence one another.

Quarks are said to come in red, green, and blue; antiquarks come in antired, antigreen, and antiblue. Each gluon has both a color and an anticolor. In building a neutron or proton, quarks constantly exchange a barrage of gluons, shifting colors with unimaginable speed. These rapid-fire adjustments are necessary because protons and neutrons must be "colorless." One way to accomplish this is to use all three quark colors to make the equivalent of white; the colors, in effect, cancel each other out. Thus, quarks must combine in trios to make neutrons and protons.

The rule governing electrical charge also requires quarks to gather in threes. Protons must have a whole electrical charge, and neutrons, none at all. Because quarks carry positive or negative charges of one-third or two-thirds that of an electron, only certain combinations of three quarks will do.

The three-step series at right depicts the exchanges involved in color conservation. The process has been greatly simplified, treating only two quarks at a time rather than the requisite three.

A single quark-gluon interaction begins as a red quark *(below, left)* and a green quark approach; the two quarks feel no mutual attraction until one emits a gluon, carrying the strong force, and the other receives it.

As soon as the red quark ejects a gluon that is red and antigreen, the quark turns green *(below, left)*. This counterbalances the gluon's antigreen, leaving one red color, just as before the gluon's birth.

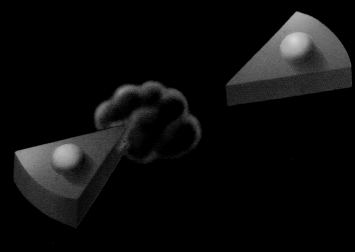

In absorbing the red-antigreen gluon, the green quark turns red *(below, right)*, its original green color canceled by the gluon's antigreen. The exchange between quarks has reversed their colors in the process of drawing them together.

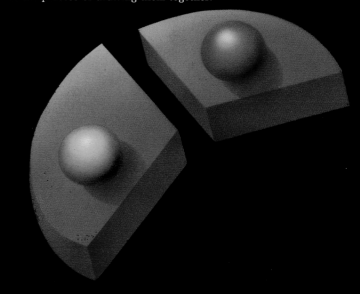

Quark Type	Mass	Charge	Color	Spin
up	310 MeV	$+\frac{2}{3}$	red, green, blue	$\frac{1}{2}$
down	310 MeV	$-\frac{1}{3}$	red, green, blue	$\frac{1}{2}$
charm	1,500 MeV	$+\frac{2}{3}$	red, green, blue	$\frac{1}{2}$
strange	500 MeV	$-\frac{1}{3}$	red, green, blue	$\frac{1}{2}$
top (?)	greater than 50,000 MeV	$+\frac{2}{3}$	red, green, blue	$\frac{1}{2}$
bottom	5,000 MeV	$-\frac{1}{3}$	red, green, blue	$\frac{1}{2}$

Counting the yet-to-be-detected version called top, quarks come in six broad types. Because each type may come in any of three colors, there are, in effect, three subspecies of each type. For convenience—and because energy is equivalent to mass—physicists express quark masses in units of energy called MeV, or millions of electron volts. A quark's electrical charge is expressed as a fraction of that of an electron. Spin, a property analogous to the angular momentum of a spinning top, is also expressed in fractions.

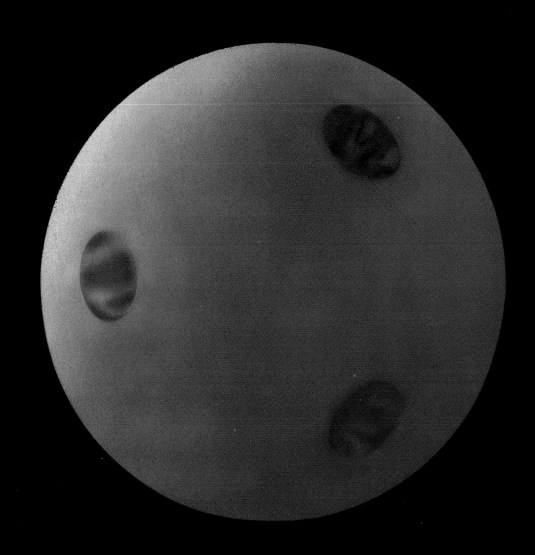

As shown in the example above, two up quarks and a down quark are needed to form a proton. Their charges of $+\frac{2}{3}$, $+\frac{2}{3}$, and $-\frac{1}{3}$ add up to the proton's required charge of 1. Their colors—red, green, and blue, respectively—cancel out to render the proton colorless. The proton's spin value of $\frac{1}{2}$ is identical to the spins of its constituent quarks.

THE VIRTUES OF VIRTUAL PARTICLES

As Werner Heisenberg pointed out more than half a century ago, the precise measurement of the properties of elementary particles is subject to considerable uncertainty. The consequences of this inherent fuzziness are rather startling: In effect, the uncertainty principle allows short-lived particles to arise spontaneously out of nothingness *(right)*.

Known as virtual particles, these evanescent objects seem to violate one of the fundamental principles of physics: the law of conservation of energy. This dictum requires that the quantity of energy in the universe remain forever constant, neither increasing nor decreasing. The law allows energy to change to or from matter, however, since matter and energy are considered equivalent; accordingly, then, matter—including elementary particles—should obey the conservation law as well. But the effects of uncertainty provide a loophole. In a tiny system, over very short time periods, energy and mass are so uncertain that they may be many times greater (or smaller) than their values a moment before. Because fluctuations for particles such as an electron-positron pair would be very brief—a few billionths of a trillionth of a second at most—they are intrinsically unobservable. As long as they pass that quickly, energy or particle excesses will never be caught in the act of violating the conservation of energy.

Despite the Alice in Wonderland logic that justifies them, a fluctuating population of virtual particles of matter can be inferred. To preserve the total electrical charge in the universe—a measure of the balance between particles and antiparticles—particles occur in matter-antimatter pairs. The electrical influence of such pairs produces measurable effects on electrons, which shift noticeably in their orbits when charged particle pairs pop into and out of existence nearby.

Uncertainty also accounts for virtual bosons, whether virtual gluons exchanged by quarks, or virtual photons transmitted between electrons. Like other virtual particles, these bosons also quickly vanish, generally being absorbed by a neighboring fermion. In the process, the force carried by the virtual boson is communicated between fermions.

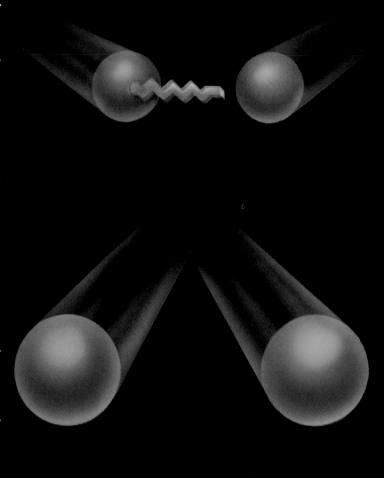

The uncertainty principle allows the electron on the left to create a virtual photon, absorbed a split second later by a neighboring electron. As a consequence of the photon's transit, the negatively charged electrons are electromagnetically repelled and move farther apart *(bottom)*.

For a fraction of a second, a virtual photon *(upper left)* violates the law of energy conservation by transforming into an electron-positron pair. A moment later, the virtual photon reappears *(lower right)* as the particle-antiparticle pair ceases to exist.

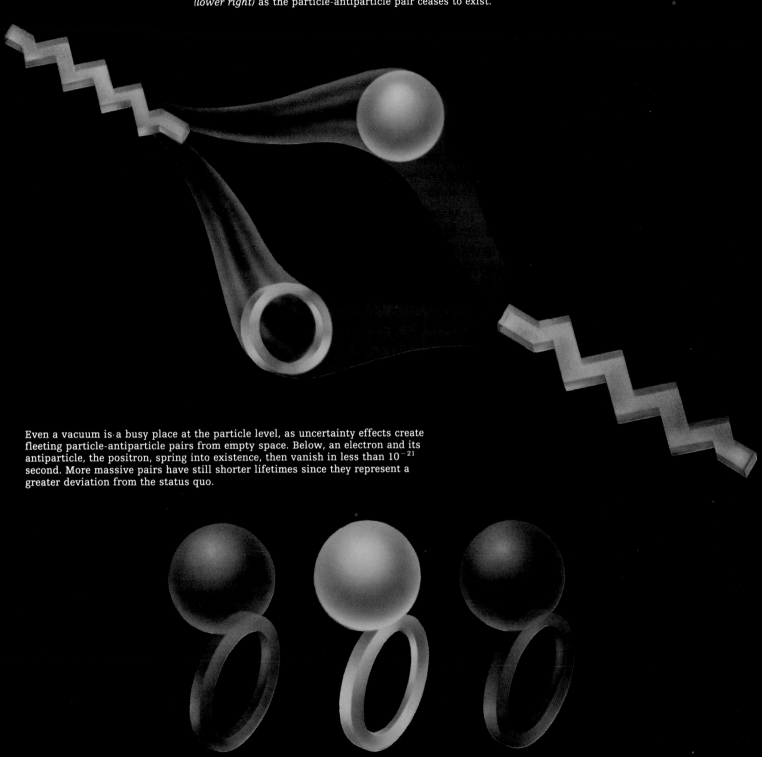

Even a vacuum is a busy place at the particle level, as uncertainty effects create fleeting particle-antiparticle pairs from empty space. Below, an electron and its antiparticle, the positron, spring into existence, then vanish in less than 10^{-21} second. More massive pairs have still shorter lifetimes since they represent a greater deviation from the status quo.

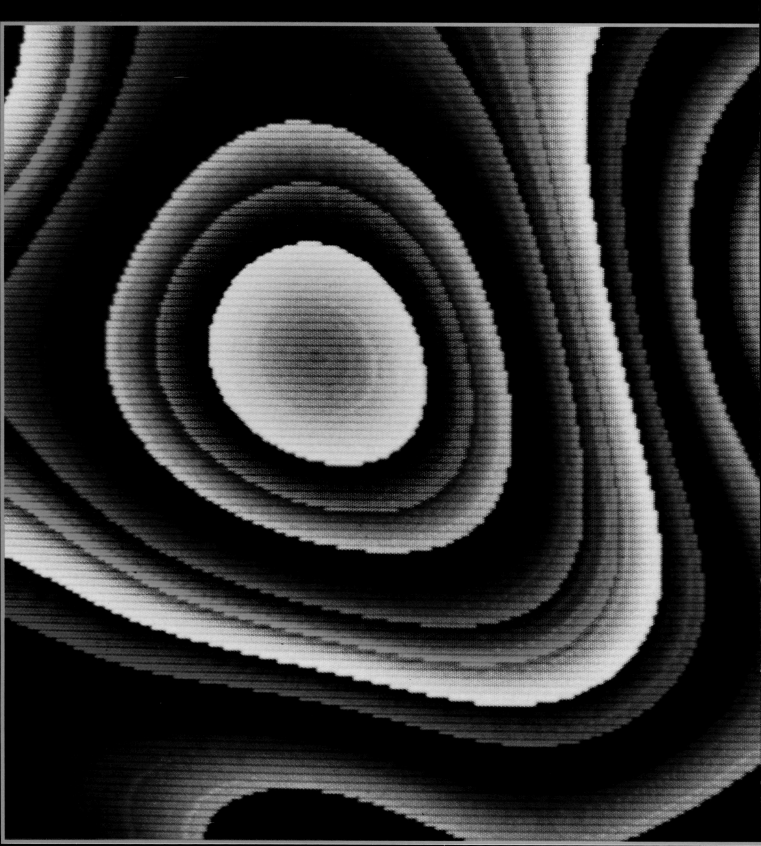

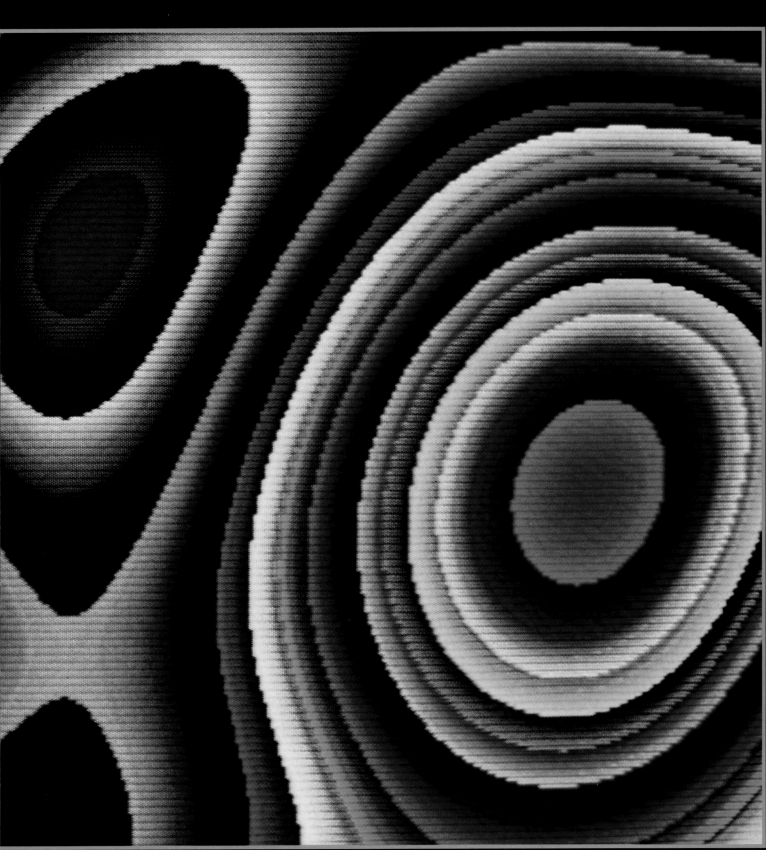

he standard model of the Big Bang painted a clear picture of the fiery epoch when protons and neutrons, jammed together in the cosmic egg, fused to form atomic nuclei. It provided succinct explanations for phenomena as diverse as the background radiation and the observed balance of elements in the universe. But it was mute on the origin of the boiling stew of particles, and hazy about the metamorphosis from clouds of hot gas to the arrays of matter that now glimmer in the cold darkness of space. Even as the Big Bang hypothesis won general acceptance in the 1960s, cosmologists began trying to fill these gaps in their understanding.

The effort was heavily influenced by discoveries about the physical processes of the universe at scales both great and small. Astronomers looking billions of light-years into space were detecting systems of galaxies whose great magnitudes and orderly structures begged for an evolutionary theory. And hints about the dynamics of the fireball began to emerge from high-energy particle research, where hordes of new particles were appearing in the detectors of huge accelerators. Such accelerators could not approach the energy levels of the Big Bang, but physicists grew increasingly confident of their ability to surmise how particles would have behaved in the stupendous temperatures and densities of the primordial explosion.

One long-sought grail of both cosmology and particle physics was the notion of what physicists called a unified field theory, a hypothesis that would use a single set of equations to describe the fields of all four forces—gravity, electromagnetism, and the strong and weak forces of the atomic nucleus. It would provide a sturdy skeleton for the body of cosmological theory by explaining, in similar terms at every scale from the subatomic to the intergalactic, all the interactions of forces and matter. Albert Einstein spent the last three decades of his life in a vain quest for a unified theory of gravity and electromagnetism, the only two forces well understood at the time. But the first real progress toward unification was made by scientists studying the forces that dominate atoms.

In the subatomic realm, governed by the rules of quantum mechanics, force and matter have a peculiar relationship, first described in the 1920s by Heisenberg, Schrödinger, Dirac, and their colleagues. At this minuscule scale, matter particles such as electrons can also be described as waves; wavelike phenomena such as photons can also be viewed as particles. Electrons change their energy level by emitting or absorbing photons, which are considered to be the carriers of electromagnetism.

Despite its success, this original theory of quantum mechanics remained incomplete. It could describe the outcome of particle interactions in general terms but could not make exact predictions about interactions between particles and radiation. When pushed beyond a rough approximation, the mathematical framework of the theory always yielded answers that were infinite or meaningless. By 1935 physicists had begun a series of fruitless efforts to devise an entirely new theory that would avoid these infinities.

The answer, as it turned out, was not a new theory but a mathematical refinement of the old. In the late 1940s, the efforts of three physicists, working separately, resulted in an improved mathematical description of particle behavior; the description was given the name quantum electrodynamics, inevitably shortened to QED. This new model used a procedure called renormalization to eliminate infinities from calculations. A sort of mathematical sleight of hand, renormalization is a delicate process that redefines the mass and charge of particles while maintaining the overall integrity of the equations. The technique worked perfectly, and soon physicists were using quantum electrodynamics to predict, with incredible accuracy, the electromagnetic interactions of such particle-accelerator products as electrons, positrons, and photons. In turn, QED suggested new approaches to the understanding of all the forces.

SOLVING QUANTUM PUZZLES

Two leaders in the exploration of physical forces were Sheldon Glashow and Steven Weinberg, who began their studies in the late 1940s as classmates at the Bronx High School of Science in New York City. The school was a fertile environment for their inquisitive minds, although they later claimed to have learned more from each other and fellow members of their science-fiction club than from teachers. The club met after school around an old laboratory table littered with Bunsen burners and test-tube racks. Discussions covered not only recent issues of magazines such as *Astounding Science Fiction* but also the latest developments in physics. Weinberg often took the lead in conversations about quantum mechanics. After the conclusion of many meetings, Weinberg and Glashow spent further hours talking about science on the telephone. "It drove our parents crazy," Weinberg remembered, "but we were sure we were going to discover wonderful things."

Both entered Cornell University, where they continued to find more excitement in debating science with each other and friends than in their courses. Glashow frequented a local poolroom, where—as he wistfully recalled—the only physics was the classical mechanics of rolling, colliding balls. The two graduated in 1954, Glashow going on to Harvard for graduate studies while Weinberg went to Princeton.

At Harvard the pace picked up for Glashow. His teachers included Julian Schwinger, one of the authors of QED. (His work on that theory would win him a share of the 1965 Nobel prize in physics.) Known for a rapid-fire speaking style and a habit of covering blackboards with scribbled equations that one student dubbed "Schwingerese," the professor paid little heed to the scheduled course matter. Instead, his students learned about what he was working on at the moment, no matter how tentative it might be. Classes sometimes began with Schwinger's admission that everything he had told them the other day was wrong. Stimulated by the professor's fearless and far-ranging intellect, Glashow asked Schwinger in 1956 to be his thesis adviser.

Schwinger gave him a problem in an obscure field, dealing with the similarity between the photon bearers of electromagnetism and the particles that carry the weak force. Trillions of times less powerful than electromagnetism, the weak force is associated with the decay of atomic nuclei and processes involving the particles called neutrinos. Schwinger's own work had led him to the idea that electromagnetism and the weak force are different manifestations of what could be, under extreme circumstances, the same force. The weak force, he suggested to Glashow, could be mediated by two proposed members of the family of particles known as bosons, which also includes photons. Schwinger called the theoretical bosons W^+ and W^-. (The signs referred to their opposite electrical charges.) It was just a hunch, but he assigned Glashow the task of checking it out.

The problem proved intractable for the young doctoral student; his thesis, submitted in 1958, left it unsolved. However, Glashow had taken the first faltering steps on an extraordinary intellectual odyssey that would end, more than a decade later, with a comprehensive theory of the common origin of electromagnetism and the weak force. During those years, some of the world's finest theoretical physicists joined the search for electroweak unification, as it came to be called. More than once their efforts were stymied by the absence of mathematical tools adequate to the task. But the possibility of ultimate success never ceased to glimmer in the distance.

The first breakthrough was Glashow's. In 1960, as he was doing postdoctoral work at the Institute for Theoretical Physics in Copenhagen, Glashow came up with a new mathematical scheme that seemed to encompass both electromagnetism and the weak force. In a paper published in 1961, he proposed a new particle, the neutral Z boson (Z^0), that would work with the two W bosons to convey the weak force in the same way photons transmit electromagnetism. In a very hot environment, he said, the W and Z bosons would be indistinguishable from photons. In other words, the weak force and electromagnetism would be unified. The key element in Glashow's hypothesis was that, unlike massless photons, the W and Z bosons must be very massive. Glashow showed that if these force carriers had a mass many times greater than a proton or a neutron, they would be effective only at very short range, which accorded with experimental data on the weak force.

Unfortunately, Glashow's equations had a flaw. He needed to use the QED theory's renormalization to rid his results of infinities, but renormalization would work only if the mass he inserted for the weak force carriers was equal

OF PARTICLES AND ENERGY

Although the mass of particles can be expressed in conventional terms such as minuscule fractions of a gram, scientists generally use another measuring stick—a unit of energy called the electron volt (eV), defined as the energy acquired by a single electron in traversing a one-volt variation of an electromagnetic field. The concept of particles as tiny bundles of energy follows from Einstein's formula for the equivalence of mass and energy, $E = mc^2$. A proton, for example, has a mass of about 10^{-24} gram, or 938,300,000 eV. The energy content of matter is significant to physicists who study the most ephemeral particles with high-powered accelerators, using the colossal machines to produce bits of matter where none existed before. These new particles coalesce from energy released when two beams of accelerated particles hit head-on. The mass of the created particles can never exceed the energy of the collisions, which is measured in billions of electron volts (expressed as gigaelectron volts, or GeV).

In today's largest accelerators, the particle beam energy reaches a few hundred GeV, just enough to create the mysterious carriers of the weak force, the W and Z bosons, whose masses are almost 100 GeV. According to contemporary quantum theory, these particles were abundant about 10^{-12} second after the beginning of expansion, when the same energy level pervaded the entire universe.

The energy levels of much earlier times are probably unattainable. During the era about 10^{-35} second after the universe began its expansion, the average energy of a particle was 10^{14} GeV. To achieve a similar level, an accelerator using the engineering incorporated in the two-mile-long, 40-GeV Stanford Linear Accelerator would have to be about one light-year long.

to zero. This would have the effect of making the weak force a long-range effect, contrary to all evidence. Glashow's proposal, generally viewed as interesting but incomplete, quickly faded into obscurity.

It languished for six years—and then was resurrected by Glashow's old classmate. Weinberg, during his graduate studies at Princeton, had also tried to develop a theory of the weak force. He maintained his interest in the subject when he took a teaching job in the physics department at the University of California in Berkeley. During the early 1960s he became interested in the work of Jeffrey Goldstone, a Cambridge University physicist who theorized that a merging of the weak force and electromagnetism should involve a new kind of boson with no mass. However, nothing matching the description of these so-called Goldstone bosons had ever showed up in an experiment. Weinberg spent a year in London working with Goldstone and another theorist, Abdus Salam, trying to eliminate the massless boson from the equations, but to no avail.

Not until 1967 did Weinberg see the way to electroweak unification. The final ingredient was the work of Peter Higgs, at the University of Edinburgh, who proposed an intricate process by which the Goldstone bosons would disappear while force-carrying particles gained mass. Weinberg realized that the Higgs mechanism could not only use up the massless bosons but also create the necessary mass of Glashow's W and Z particles. By combining the Glashow, Goldstone, and Higgs equations, he was able to predict the mass of the W bosons at eighty-three billion electron volts (GeV)—more than eighty times the mass of a proton. The Z^0 was more massive still—ninety-three GeV. He announced his theory with a brief paper in the November 1967 issue of the *Physical Review Letters* but was unable to wade through the immensely complex renormalization calculations that would finalize his breakthrough. As a result, the paper received little attention and was soon forgotten.

Meanwhile, Abdus Salam, Weinberg's erstwhile coworker in London, was developing a similar theory. A leading expert on the weak force, Salam had risen to scientific eminence from humble origins in British-ruled India by dint of his extraordinary mathematical ability. He began graduate work at Cambridge University in 1946, at the age of twenty, and went on to win a reputation as a brilliant theoretical physicist. Salam often crossed paths with Glashow and Weinberg—sometimes less than happily. In 1959, at a lecture in London where Glashow outlined his ideas on electroweak unification, Salam took umbrage. "Here was this slip of a boy," he later recalled, "claiming that he had solved the problem I had been wrestling with for months!" To the younger man's great chagrin, Salam pointed out inconsistencies in the mathematics Glashow was using to describe unification. An unfortunate consequence of the encounter was that Salam lost respect for Glashow's work, and ignored it for years to come.

In 1964 Salam arrived, by a different mathematical route, at a hypothesis that used many of the ideas in Glashow's 1961 paper. Like Glashow's, his theory was not susceptible to renormalization; in its incomplete state, it too attracted little attention. In 1967, after learning about the Higgs mechanism, Salam had an insight similar to Weinberg's. He worked out the theory and introduced the idea during lectures at the Imperial College in London just as Weinberg was preparing his own version for publication. But the mathematics of renormalization proved to be as daunting for Salam as they had for Weinberg; he soon abandoned the effort and moved on to other work.

Finally, in 1971, a Dutch graduate student named Gerard t'Hooft conquered the beast, coming up with a clever mathematical approach that made renormalization possible for both Weinberg's and Salam's theories. Word of the method quickly spread among particle physicists, and interest in the subject soared. Weinberg himself returned to the topic with new spirit, publishing three important articles in late 1971. Other scientists joined the effort: Weinberg's 1967 paper, which had been cited in scholarly publications just once in 1970 and four times in 1971, was mentioned 64 times in 1972 and 162 in 1973. Snatched from the shadows, the theory of electroweak unification finally had enough substance to attract the interest of experimental physicists.

Physical verification promised to be hard to come by. Detection of a Z^0 boson would be strong evidence, but the accelerators of the early 1970s had only a fraction of the power required to produce such a massive particle. Weinberg therefore began to study indirect approaches. His theory predicted the existence of a neutral current in the interaction of neutrinos with protons or neutrons (collectively called nucleons). In most such events, the weak force would be transmitted by a charged W boson, which would cause the nucleon to decay and transform the neutrino into a charged particle called a muon. However, sometimes the force carrier would be a Z^0, which would also cause the nucleon to decay but leave the neutrino unchanged. If a beam of neutrinos is focused on a particle detector, where neutral particles like the neutrino or the Z^0 are invisible, the signature of the neutral current would be an apparently spontaneous shower of charged particles that did not include a muon.

The problem was to witness enough of these interactions to prove the existence of a neutral current. Neutrinos are small, fast, influenced only by

On stage at the Nobel ceremony in Stockholm in 1979, American physicists Sheldon Glashow *(left)* and Steven Weinberg *(right)* flank Abdus Salam, in formal Pakistani dress. The three shared the physics prize for linking electromagnetism to the weak nuclear force, which governs radioactive decay.

the weak force, and very unlikely to interact with anything. One of the world's largest accelerators, at the European Center for Nuclear Research (CERN), near Geneva, Switzerland, routinely poured a billion neutrinos per second into its huge particle detector; collisions occurred only about once per minute. Nevertheless, as part of a major effort to detect quarks (the building blocks of nucleons), CERN researchers had compiled more than half a million photographs of neutrino events by late 1971. As results of the quark hunt continued to roll in, a small team of researchers led by physicist Paul Musset began combing through the records. They looked for the characteristic traces of the neutral current and tried to determine their true origins.

By the summer of 1973, they had found about a hundred neutrino-nucleon events from which no muon emerged, enough to convince Musset. He was bursting with the news when he attended a conference of particle physicists that fall in Aix-en-Provence, France. Driving through the cobblestoned streets of the ancient town, he spied a figure he thought he recognized trudging from the train station with a heavy suitcase. "Are you Salam?" he asked the man, who was in fact Abdus Salam. "Get in the car. I have news for you. We have found neutral currents." Salam later recalled that his pleasure at these tidings was doubled by the prospect of not carrying his luggage another step.

Weinberg, at the same conference, was less sure that the theory was proved, but the next year the CERN results were duplicated by several other research groups. By the end of 1974, little doubt remained about neutral currents, and the Weinberg-Salam model was well on its way to acceptance. Glashow, who had laid the foundation for the theory, found himself excluded from the credit, a circumstance that rankled even more deeply as it became clear that the electroweak unification might win its authors a Nobel prize. Happily, the threat to his long friendship with Weinberg was removed by the Nobel committee, which in 1979 awarded the physics prize in equal shares to Glashow, Weinberg, and Salam. And in 1983, physicists at CERN crowned the theory with the detection of actual W^+, W^-, and Z^0 particles.

A GLIMPSE OF THE BEGINNING

For cosmologists, electroweak unification meant that the clock could be turned back to a time just one-trillionth of a second after the beginning of expansion, when the fireball had a temperature of 10^{16} degrees Kelvin (millions of times hotter than the center of the Sun) and was still so dense that a thimbleful weighed 100 million tons. The next challenge was clear. Now that the theorists understood the interactions of particles and forces of that time, they focused on an even earlier period, when the strong force was also unified with the electroweak. The object of their quest was a Grand Unification Theory, more simply known as a GUT, that would bring together all the forces except gravity.

One of the first theorists to attack the grand unification problem was Sheldon Glashow, who began his assault even before CERN's detection of the neutral current confirmed the electroweak hypothesis. After his electroweak ruminations, Glashow had gone on to become an expert on the strong force. By 1972 it was understood that principles similar to quantum electrodynamics governed the interaction of quarks in nucleons, mediated by the strong force. The name of the strong force theory, quantum chromodynamics (QCD), consciously mirrored QED. However, it was much more complex. While electromagnetism is carried by a single photon, quantum chromodynamics called for no fewer than eight gluons, as the bosons carrying the strong force are called. And those eight bosons promised to put the mathematics of a GUT into a whole new category of difficulty.

With Howard Georgi, a Harvard postgraduate fellow, Glashow initiated hectic daily work sessions in his office at Harvard's Lyman Hall. Every morning he would barrage Georgi with new ideas about the strong force. He expected his colleague to try to punch holes in each hypothesis; if Georgi could not find a flaw right away, that subject would be the focus of his attention for the day.

One autumn afternoon in 1973, the two wrangled for hours about approaches to a GUT, coming to no conclusion. After dinner, Georgi began to follow up the leads that had emerged during the afternoon's debate. Much to his surprise, he soon produced a mathematical model that seemed to fit the facts, uniting the strong and electroweak forces. "I was very excited," he remembered later. "I sat down, had a glass of scotch, and thought about it for a while." Georgi then approached the problem another way, and this effort also turned out successfully. "So I got even more excited and had another scotch."

Like the electroweak unification theory, Georgi's model involved changes in one kind of particle that were precisely offset by changes in other particles. However, the energy levels required were 10 trillion times higher—equal to

THE NUMBERS OF COSMOLOGY

The very large and very small quantities common to cosmology are usually represented as powers of ten. The diameter of an atom, about .00000001 centimeter, is expressed as 10^{-8} centimeter; the exponent (-8) signifies the number of decimal places in the fraction. Similarly, the Sun's approximate mass in kilograms is represented as 10^{30}, which in ordinary decimal notation would be a one followed by thirty zeros. This numbering system not only is concise but also allows widely divergent quantities to be easily compared by adding or subtracting exponents instead of performing tedious divisions and multiplications.

those that existed at 10^{-36} second after the beginning of expansion. In that environment, gluons, normally the carriers of the strong force, would be the equivalent of photons and the W and Z bosons. Quarks and electrons would also be interchangeable, in transactions mediated by an entirely new family of twelve extremely massive particles, later lumped under the name X bosons.

Georgi's private celebration was dampened when he discovered a startling aspect of his hypothesis. Working through the mathematics, he realized that the equations allowed—in fact, dictated—the eventual decay of every proton by the transformation of one of its quarks into an electron or a positron. If this theory about the very earliest epoch of the universe was correct, it also foreshadowed a possible end. When protons evaporate, the universe must become nothing but cold, dark space, devoid of even the atoms that now make up planets and stars.

So it was with mixed emotions that Georgi showed his solution to Glashow the next morning. Glashow did not let the fate of protons temper his pleasure at what seemed to be a genuine breakthrough. "It wasn't shattering," he later recalled. "We know the sun will burn out in a few billion years. That matter falls apart a long, long time afterwards is scarcely an upsetting idea."

Georgi and Glashow published their Grand Unification Theory in February 1974. It won adherents—and also a slew of imitators, as other researchers followed the same line of attack to develop similar theories. All predicted the X bosons of the GUT Era, as the time around 10^{-35} second came to be called. But how could these mathematical visions ever be verified? No particle accelerator could ever achieve the energy levels required to produce the X bosons. The only possible experimental evidence for the GUTs would be to catch a proton in the act of decaying. Yet protons are unimaginably long-lived. The best estimates set their average life at about 10^{31} years, more than a billion billion times the present age of the universe.

Obviously scientists could not watch one proton long enough to see it decay. But because the longevity of the particles was an average figure, with some protons enjoying a far longer life span than others, decay might be observed if a great many protons were watched. One early attempt to do this, beginning in 1981, used a vast tank filled with 8,000 tons of purified water, located in a half-mile-deep salt mine near Cleveland, Ohio. The water contained 10^{34} protons, more or less, and the underground location protected it from spurious radiation that might give a false indication of proton decay. In the act of disintegration, a proton should produce a minute flash of light, detectable by sensitive phototubes lining the walls of the tank. One scientist captured the ambivalence of his fellows in a toast at the opening of the facility: "To the

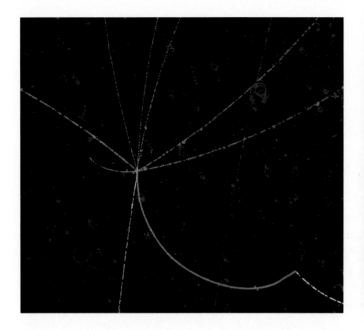 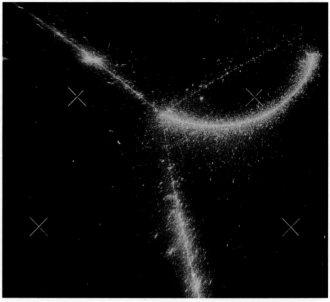

proton—may it live forever! But if it has to die, let it die in our arms!" The subject of this toast has not yet obliged. No protons have been seen to die in Ohio, or in similar experiments conducted in Japan, India, and Italy. But the watch continues.

For cosmologists, the predicted demise of matter particles was less important than its corollary in the GUT models: the creation of matter from the radiation that pervaded the universe in the GUT Era. In that high-energy environment, massive X bosons would supposedly decay into showers of particles that included quarks and electrons. As the expanding universe cooled, these would eventually combine to form atoms. The theories even predicted the resulting proportion of matter to energy in the universe, a ratio that closely matched the balance between the observed density of matter and the intensity of the background radiation. The GUTs fared well on the proving ground of the cosmos.

SOME NAGGING PROBLEMS

Theoretical advances toward the unification of forces did more for cosmology than show how matter could have been created and may ultimately dissolve. The grand unification theories also seemed to open new approaches to problems posed by the standard model of the Big Bang. Despite its overall triumph as an explanation of cosmic origins, that model left unanswered some riddles about why the universe appears the way it does today.

One such difficulty, known as the horizon problem, was raised by the apparent uniformity of the universe, as indicated by the temperature of the cosmic background radiation. This varies by less than one part in 10,000 from one area of the sky to another. All parts of the universe must have emerged from the primordial fireball with virtually identical temperatures—a circumstance that required an exchange of radiation between regions to smooth out any initial thermal imbalance. According to the standard model, however, some segments of the expanding cosmos had always been beyond each other's

Bubble chamber. In the false-color photo at far left, the clash of a proton and its antimatter equivalent, an antiproton, registers in the liquid hydrogen of a bubble chamber, where the antiproton's path shows up in blue. The collision produces eight short-lived particles: four negative pions *(green)* and four positive pions *(red)*, one of which decays into a muon *(yellow)*.

Streamer chamber. Above, particles from a similar collision ionize helium in the electric field of a streamer chamber to leave luminous trails, or streamers. An antiproton angling in from below strikes a helium proton, causing pions to shoot toward the upper corners while a nucleus of tritium, a form of hydrogen, arcs right.

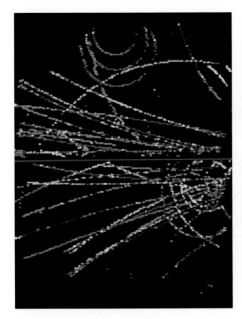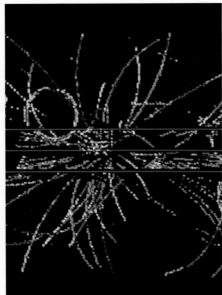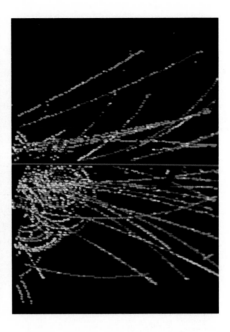

Electronic chamber. A sectioned electronic chamber filled with ethane and argon gas senses the passage of subatomic particles with thousands of tiny wires. Here, an antiproton and a proton speed in from either side to collide with each other head-on. The resulting spray of particles is color-coded to indicate direction.

cosmic horizon, separated by a distance greater than light could traverse during the life of the universe. These regions could never have been in touch with each other.

Another, related puzzle was the existence of galaxies. These clumps of matter were hard to reconcile with the smoothness of the background radiation. If the universe was homogeneous in its infancy and has expanded evenly since then, every particle of matter should have been carried farther and farther away from every other particle; no areas of higher density would have formed to act as gravitational seeds for galaxies.

But one question more than any other spurred research that led to a new model of how the cosmos evolved. Since the earliest days of relativistic cosmology, astronomers had been perplexed by the fact that the universe appears to be nearly flat, balanced between expanding forever and gravitationally collapsing upon itself. As Alexander Friedmann had shown in the 1920s, the average density of matter defines the curvature of space-time, making the difference between an open universe that will expand forever and a closed universe that will eventually contract. Only if the actual density is equal to a critical density, derived from the field equations of relativity, can the universe be flat. Both values diminish with the expansion of the universe; the present critical density is about three hydrogen atoms per cubic yard. Most calculations based on observation showed the actual density and critical value differing by no more than a factor of 100, even though there was no clear reason why it should not be a million times greater or less.

The flatness of the universe today, as one cosmologist pointed out, seemed less likely than a pencil balancing perfectly on its point after millions of years. In 1979 the magnitude of this improbability was given mathematical expression when Princeton astronomers Robert Dicke and James Peebles (the same pair who were involved in discovering the background radiation) calculated just how flat the early cosmos must have been. One second after the moment of creation, the density of the universe was equal to the critical value within one part in a million billion; at the end of the GUT Era, the difference between the two was a decimal fraction with forty-nine zeros. Such agreement could not possibly be a result of chance. Somehow, Dicke and Peebles declared, the laws of physics must require the universe to emerge from the Big Bang extremely flat.

In the spring of 1979, Dicke spoke at Cornell University on the extraordinary cosmic balance between galloping expansion and cataclysmic collapse. One fascinated listener, a young particle physicist named Alan Guth, took the problem very much to heart. By the end of the year, he would find an answer—and a completely new way of viewing the Big Bang.

AN EXPANSIVE SOLUTION

While growing up in New Jersey, Guth had amused himself by designing rockets and calculating how high they would take astronauts. Later, when he attended the Massachusetts Institute of Technology, he was drawn to more theoretical forms of physics. He earned a Ph.D. at MIT in 1972, then spent several years in particle research at Princeton, Columbia, and finally Cornell. There, he began to focus on cosmology. The catalyst was a visit to the university by Steven Weinberg, who convinced Guth that the Big Bang was a good laboratory for GUTs.

In the fall of 1979, Guth took a year's leave from Cornell to work at the Stanford Linear Accelerator Center in California. Ideas he had picked up from his cosmological studies began to mix with what he knew about high-energy physics, and the seed planted by Dicke's lecture took root.

Guth was working on the idea that the separation of the strong force from the electroweak at the end of the GUT Era—a process known as a phase transition—could be similar to the change of water into ice. His first assumption was that the phase transition would take place as soon as the universe cooled to a critical temperature, just as water usually freezes at thirty-two degrees Fahrenheit. But as he worked on the problem, Guth remembered the phenomenon called supercooling: If water is cooled very rapidly, it can remain liquid well below its normal freezing temperature, then suddenly freeze all at once. He began to speculate about the effects on the early universe if the phase transition between the strong and electroweak forces occurred only after a significant amount of supercooling.

On Thursday, December 6, 1979, in the course of a long conversation with a visitor about GUTs, the idea of creating matter from X bosons came up. It stuck in Guth's mind, and he sat down in his office with a notebook, trying to work out a mathematical expression of his thoughts. At home that night he kept writing, long after his wife had gone to bed, filling page after page of the notebook. In the small hours of Friday morning, he had a flash of insight: Supercooling in the cosmic fireball provided a perfect explanation for Dicke's flatness paradox.

"I broke my bicycle record getting to work next morning," Guth recalled eight years later. "That's when I actually worked out all the equations. I wasn't sure whether my ideas would turn out to be right or not. Still, it made me feel a little queasy inventing something that dramatic."

Guth's version of the nascent cosmos was revolutionary: At 10^{-35} second, as the still-tiny universe cooled beyond the temperature at which the strong and electroweak forces should have separated, it passed instead into a supercooled state. This created a peculiar condition known as a false vacuum, where, according to the field equations of general relativity, gravity pushes matter apart instead of drawing it together. In the span of about 10^{-32} second, the gravitational repulsion caused the universe to double in size 150 times. Much smaller than a proton at the beginning of this inflation, the cosmos ballooned to a diameter of about ten centimeters. A grain of sand swelling in similar proportion would grow far larger than the visible universe.

Inflation stopped, Guth maintained, when the phase transition began, converting the energy of the false vacuum into particles through the Higgs mechanism. The grapefruit-size cosmos continued to grow, but now gravity had returned to its normal role, and under its influence the rate of expansion of the new matter began to slow. After this point, Guth's theory agreed with the standard model of the Big Bang.

The hypothesis gave logical answers to some of the vexing questions of the standard model, such as the horizon problem. Before inflation, what is now the visible universe was so small that regions that are today on opposite sides of the universe, as viewed from Earth, were in direct contact with each other. Within the tiny cosmic seed, there was time for energy to become evenly distributed through every part. Space then inflated exponentially, at a rate far in excess of that envisioned by the standard model. This growth would have been limited by the speed of light if matter had been moving through space, but the expansion of space itself is under no such constraint. Separated by inflation at many times the speed of light, the once-adjacent regions were never again in communication.

Inflation solved the flatness problem by reducing it to a simple exercise in four-dimensional geometry. No matter what curvature the universe originally had, it would have been flattened by its exponential expansion, in the same way that in three dimensions the surface of a balloon becomes less and less curved as it is blown up. After space had doubled in size 150 times in a minuscule fraction of a second, it became virtually indistinguishable from a flat universe. Its average density, which is proportional to the curvature, would therefore be at or very near the critical value.

Almost as soon as he finished his calculations, however, Guth saw a major flaw in his model. The transition from a supercooled state would not have taken place simultaneously throughout space but at differ-

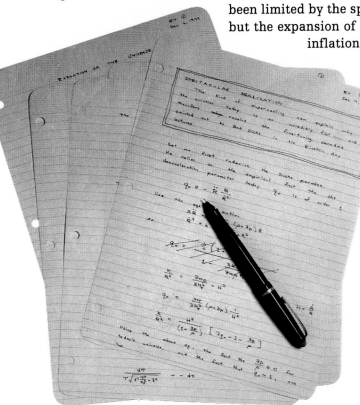

One night in late 1979, after filling pages of his notebook with calculations, physicist Alan Guth jotted the words SPECTACULAR REALIZATION and proceeded to describe a new theory of cosmic birth. According to Guth's scheme, the primeval universe expanded by at least 10^{25} times in the merest fraction of a second, a drastic inflation during which most matter and energy were forged.

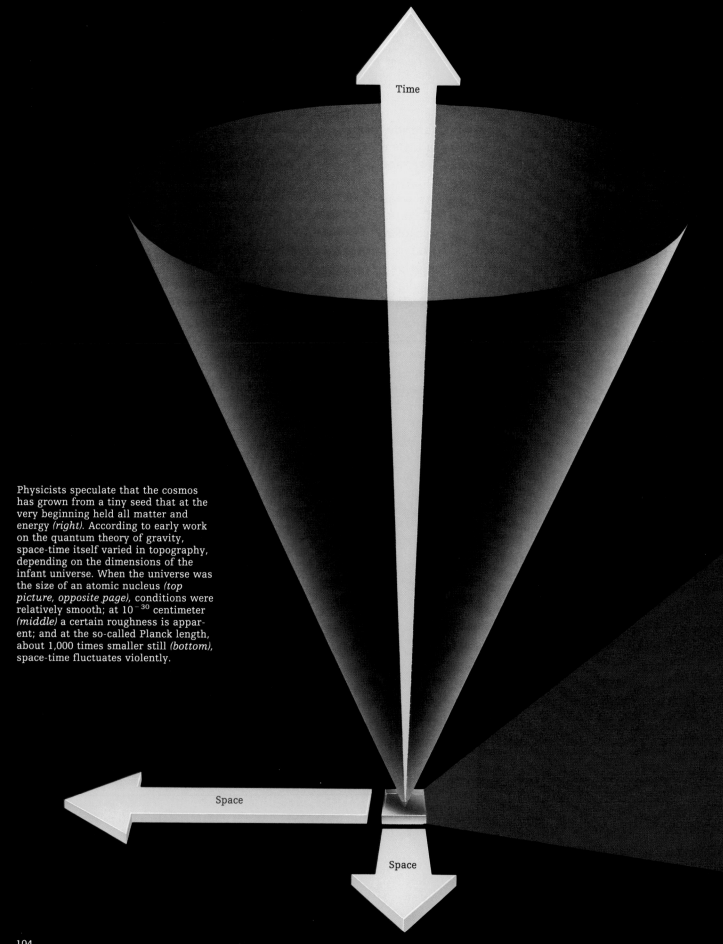

Physicists speculate that the cosmos has grown from a tiny seed that at the very beginning held all matter and energy *(right)*. According to early work on the quantum theory of gravity, space-time itself varied in topography, depending on the dimensions of the infant universe. When the universe was the size of an atomic nucleus *(top picture, opposite page)*, conditions were relatively smooth; at 10^{-30} centimeter *(middle)* a certain roughness is apparent; and at the so-called Planck length, about 1,000 times smaller still *(bottom)*, space-time fluctuates violently.

Time

Space

Space

SPECULATIONS ON THE PLANCK ERA

To understand the structure of the universe on the grandest scales, cosmologists must track backward through time, pressing their theories toward the moment when everything came to be. With the advent of so-called grand unification theories in physics, they have modeled events and conditions in the early universe almost all the way back to the beginning. The point beyond which they cannot yet go—at least with current theories—is 10^{-43} second after the Big Bang *(pages 119-131)*, a moment known as the Planck Time after German physicist Max Planck.

This barrier exists because prior to the Planck Time, during the period called the Planck Era, all four of the known fundamental forces of nature are presumed to have been indistinguishable or, in effect, unified. Although physicists have devised quantum theories that unite three of the forces, one by one, through eras going back to the Planck Time, they have so far not been able to bring gravity into the fold. To do so, they must reconcile the laws of quantum mechanics, which operate at microcosmic scales, and Einstein's theory of gravitation *(pages 48-57)*.

Preliminary work on a quantum theory of gravity has pointed out a number of startling implications. For one, quantum gravity would depend on the universe's having more than four dimensions. Furthermore, during the Planck Era, quantum gravity and the universe may have been one and the same—an infinitesimally tiny quantum object. At this scale, spacetime itself would be subject to unpredictable fluctuations analogous to the ones that cause particles to pop into and out of existence in ordinary space-time *(pages 88-89)*. This notion has led theorists to describe the Planck Era universe as a kind of writhing foam.

10^{-12}cm

10^{-30}cm

10^{-33}cm

BUBBLING OUT OF NOTHINGNESS

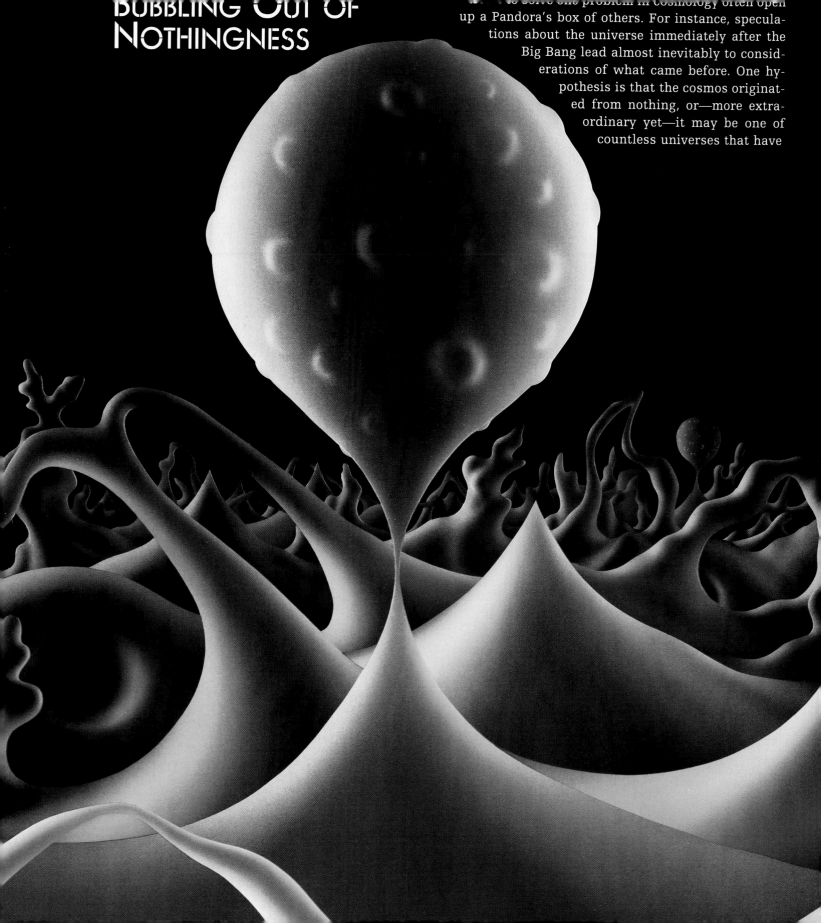

...s solve one problem in cosmology often open up a Pandora's box of others. For instance, speculations about the universe immediately after the Big Bang lead almost inevitably to considerations of what came before. One hypothesis is that the cosmos originated from nothing, or—more extraordinary yet—it may be one of countless universes that have materialized out of empty space. The key to such assertions lies in the nature of "nothingness." In everyday understanding, nothingness is empty, a vacuum. But to quantum physicists, a vacuum is something else again, an inherently unstable condition, ripe with latent energy, in which neither space nor time in the classical sense exists.

According to some theorists, the nothingness that precedes space and time may have been the same kind of fluctuating foam as that of the Planck Era (*page 105*). Perhaps engendered by the vibrations of proto-particles called superstrings *(below)*, these vacuum fluctuations might be visualized as tiny bubbles, like those shown here. Some bubbles would simply appear and disappear, but others might suddenly expand into a whole cosmos. In theory, then, innumerable alternate universes, each a separate bubble, might exist next door, unreachable from our own space-time.

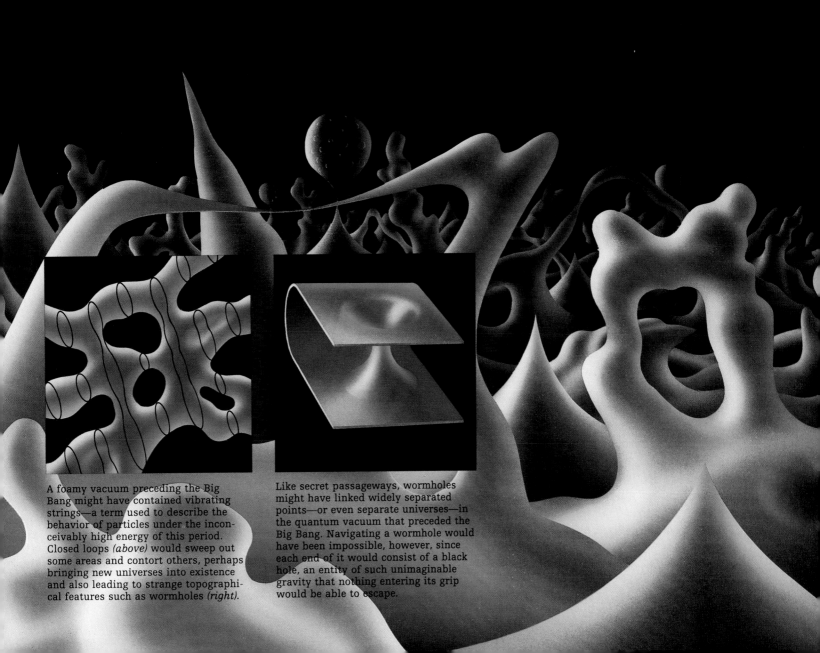

A foamy vacuum preceding the Big Bang might have contained vibrating strings—a term used to describe the behavior of particles under the inconceivably high energy of this period. Closed loops *(above)* would sweep out some areas and contort others, perhaps bringing new universes into existence and also leading to strange topographical features such as wormholes *(right)*.

Like secret passageways, wormholes might have linked widely separated points—or even separate universes—in the quantum vacuum that preceded the Big Bang. Navigating a wormhole would have been impossible, however, since each end of it would consist of a black hole, an entity of such unimaginable gravity that nothing entering its grip would be able to escape.

ent times in different places. The result would be a universe with many bubbles of space-time, each with different laws of physics caused by the different ways that the phase transition occurred within them. The bubbles would have formed clusters, like huge bunches of grapes. Boundaries between the bubbles in a cluster would have been clearly detectable. But this is definitely not what astronomers see when they survey the heavens.

Guth worked on his theory for months, then decided to publish it, warts and all, in the hope that another scientist would be able to solve the bubble problem. His wish was soon granted. Andrei Linde, a young high-energy physicist at the Lebedev Institute in Moscow, thought he saw a connection between Guth's theory and his own work on the Higgs mechanism. He plunged into a rescue effort but made little progress for months. He later recalled that his frustration made him feel physically ill: "I could see no way to improve the situation and I could not believe that God would miss such a good possibility to simplify the creation of the universe."

The summer of 1981 brought a breakthrough. Linde found that if the phase transitions in the different parts of space-time had occurred somewhat more gradually than in Guth's theory, the result would have been a universe free of bubbles and boundaries. In the Linde model, the details of the phase transition still varied from region to region, but the result was a smooth fabric of adjacent cells of space-time, called domains. Each would have developed independently, growing much larger than a present-day galaxy by the end of the inflationary era. What is now the visible universe, then just ten centimeters in diameter, would have been dwarfed inside a single domain. After billions of years of expansion, the domain boundaries would be far beyond the range of observation. Our universe, with its uncounted billions of galaxies spanning more than 30 billion light-years, is only a billion-trillionth part of a domain, and that domain is but one among an unknowable number.

Other cosmologists first got wind of Linde's hypothesis when he presented it in October at an international seminar in Moscow. In April 1982 two American researchers, Andreas Albrecht and Paul Steinhardt, published similar conclusions based on independent work. Even as the new inflationary model gained currency, however, Linde and others were beginning to question it. In 1983 Linde proposed yet another model that eliminated supercooling and the phase transition altogether, replacing them with earlier phenomena that produced similar effects. This hypothesis and others became the basis for a debate that continues into the late 1980s. The controversy is limited to the mechanism of the inflation, however; little doubt remains that exponential expansion during the Big Bang created the cosmos we see today, as well as an immeasurably larger universe forever beyond our view.

INFLATION PAYS OFF

Astronomers and physicists were still studying the complex mathematics of the new inflationary model when a group of theorists stumbled upon an unexpected bonus: The process of inflation not only eliminated the troublesome bubble intersections but also seemed to produce the necessary precursors to the large-scale structures of the modern universe.

In the summer of 1982, a number of leading cosmologists gathered at Cambridge University for a special event called the Nuffield Workshop on the Very Early Universe. Among them were Guth and Steinhardt, who took part in discussions of galaxy formation in an inflationary cosmos. Things got off to a rocky start, since the theorists were concerned that inflation, by smoothing out the inhomogeneities of the early universe, would not have left enough density fluctuations to produce the structures of matter observed in the cosmos today. Then one participant pointed out that the situation might be saved by applying quantum theory.

The quantum world is ruled by probabilities. For example, the position of an electron cannot be described precisely; all that can be stated is the probability of its existing at a certain location at a given instant. Since this uncertainty has little impact on the world beyond the scale of atoms and molecules, cosmologists had not thought to consider its influence on cosmic structures measured in light-years. The insight at the workshop was that, in an inflationary scenario, the entire vastness of space was once smaller than an atomic nucleus. In so tiny a region, quantum effects—probabilistic particles flickering into and out of existence—could easily produce density variations that inflation later stretched to the sizes of galaxies and clusters.

Excited by this new concept, the Nuffield cosmologists broke into working groups to develop the mathematics that would plug quantum fluctuations into the inflation theory. After three weeks of discussion and disagreement they settled on an answer that looked surprisingly familiar. The resulting clumpiness resembled the contours of the post-Big Bang universe predicted almost a decade earlier by Soviet cosmologist Yakov B. Zel'dovich. It was an auspicious match, because Zel'dovich's model provided the best explanation yet for the current shape and structure of the cosmos.

Zel'dovich was of a different breed from the whiz-kid particle physicists who had recently brought so much to cosmology. He was still a sparsely educated teenager when he found his first job in science as a technician at the Institute of Processing of Useful Ores in Leningrad. One day in 1931, on an excursion to the Leningrad Physical-Technical Institute, he fell into a discussion with staff members on the crystallization of nitroglycerin. Impressed by the clever youth, they invited him to work in their lab in his free time, and soon Zel'dovich transferred to the institute officially. In 1936, at the age of twenty-two and without the benefit of university coursework, he wrote and successfully defended his dissertation for the degree of Candidate of Sciences, similar to a doctorate at a Western university.

Although chemistry would remain a lifelong interest, Zel'dovich was always attracted to the grandest and most fundamental problems in physics. As Soviet scientists began to investigate nuclear fission in the 1940s, he did pioneering work on the theory of explosive and controlled chain reactions and became a central figure in Soviet fusion research. In the mid-1950s he entered the field of elementary particle physics, where he made significant theoretical contributions to the understanding of the weak force. And by the 1960s Zel'dovich, like so many others, had turned toward the heavens as the ultimate physics laboratory.

He soon moved to the front rank of astrophysicists, publishing important papers on subjects such as neutron stars and black holes. Then he found himself drawn to a puzzle of a larger order: the evolutionary links between the Big Bang and the observable universe of today.

During the 1950s and 1960s, observatories around the world had reported cosmic structures of previously unimagined size. Astronomers mapped clusters of thousands of galaxies, then began to discern superclusters, colossal aggregates of as many as a dozen clusters, with a total mass of 100,000 galaxies. It was generally acknowledged that such uneven matter distribution must have originated in the earliest epoch, but no one could describe with precision the sequence of events that led from that initial clumpiness to the huge configurations of galaxies. Zel'dovich found the challenge irresistible.

Like others who had pondered the problem before him, he assumed that clumpiness in the early universe must have taken the form of slight compressions and rarefactions of matter and energy from region to region. As the universe expanded, gravity would have pulled more matter into the compressed parts, forming clouds of gas from which stars, galaxies, and larger structures could later coalesce. The problem was one of balance. The early fluctuations must have been small, or they would have caused ripples in the background radiation; no such ripples had been observed. But the density variations must also have been large enough to grow into the biggest structures now visible in the heavens.

After years of work, Zel'dovich developed a model that reconciled these contradictory elements. The final version of his theory, published in 1980, held that neutrinos "free-streamed" throughout the early universe, flowing in every direction at such high velocities that they would have blown away all fluctuations less massive than the approximate mass seen in a supercluster. Zel'dovich calculated that gravity would have caused these compressed regions, containing about seven-eighths of all matter, to collapse into vast irregular gas clouds that looked like flattened pancakes. The remaining matter would disperse throughout the volume of space vacated by the condensing pancakes. The result would be a cellular structure of thin walls enclosing huge voids, from 100 to 400 million light-years across and nearly empty of matter. As eons passed, the compressed matter of the pancakes would break down first into cluster-size clouds, then into galaxies and their stars.

Zel'dovich's model did a good job of explaining the structures that astronomers were mapping. But because it described only the shape of the initial density fluctuations and not their origin, it remained just one among several competing evolutionary theories. Then, in 1982, Guth and his colleagues at the Nuffield workshop discovered the resemblance between the clumpiness predicted by the pancake theory and the variations in density produced by quantum fluctuations in the cosmic seed. Cosmologists seemed tantalizingly close to a model that could explain the history of the universe, step by step, from the GUT Era to the present day.

But Zel'dovich's theory had a weak link: the free-streaming neutrinos. True, the Big Bang would have produced these particles in the necessary abundance—about a billion for each proton or neutron. But neutrinos were generally presumed to have no mass whatever, and to smooth out density fluctuations they would have to have some. Assigning neutrinos even the tiny mass needed to make the model work—about twenty electron volts, or 1/25,000 the mass of an electron—flew in the face of accepted particle theory.

Nonetheless, there was some experimental support for Zel'dovich's proposal. Measuring the mass of neutrinos is not easy, because the elusive particles almost never interact with other matter. A neutrino can hit Texas, pass through Earth, and come out in Australia without leaving a trace. Occasionally, however, one hits a proton in the nucleus of an atom, causing the proton to decay into detectable particles. The experiments set up to catch protons in spontaneous disintegration frequently detect such neutrino-initiated proton decay; similar detectors have been used since the 1960s to study neutrinos generated by the Sun. By the mid-1980s physicists had ac-

A map of one region of the sky reveals galaxies gathered around vast, bubble-shaped voids. Measuring more than 100 million light-years across, the voids may be the products of ancient explosions.

A Cosmos Structured by Strings

One of the most intriguing theories in physics ascribes the observed large-scale structure of the universe to the influence of cosmic strings—remnants of the early cosmos that are loop shaped and so massive that a snippet an inch long and a billion-billionth the thickness of a proton would weigh as much as a mountain range. These hypothetical objects are thought to have been created during so-called phase transitions, critical periods when the universe underwent a change analogous to the way water turns to ice or vapor.

The first phase transition occurred a minuscule fraction of a second after the Big Bang. As the infant universe cooled, it went from a state of pure energy to one of energy and matter: In effect, matter condensed into existence, and during other transitions, similar processes separated forces such as the strong and weak nuclear forces from one another. At each stage, uneven transitions might have created flaws in space-time, just as uneven freezing produces cracks in ice. Within these defects, space-time would retain the forces and matter of the previous phase.

If they exist, cosmic strings would have neither beginning nor end. They would form loops or stretch to infinity, vibrating with a rhythm that would send gravity waves rippling through space. Because short strings would oscillate rapidly, dissipating their energy in a few million years, only the longest strings, with ponderous rates of oscillation, would still be around. But as illustrated on the next two pages, the long-gone short strings may have been the prime reason for the clustering of galaxies seen today.

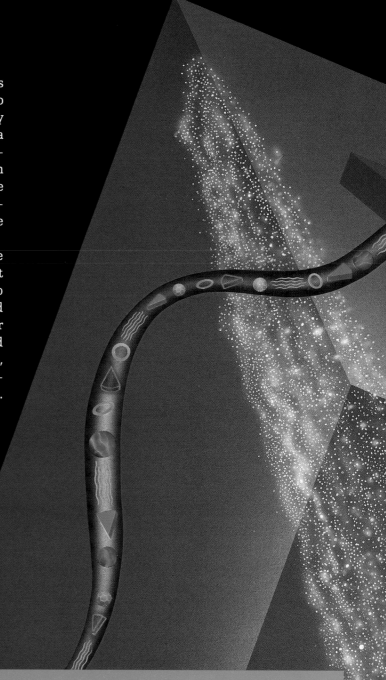

This computer-generated sequence tracks a cosmic string through one oscillation cycle. Whatever its size, a string executes about 10,000 cycles before running out of energy; thus, large, slow-moving strings live much longer than rapidly vibrating short ones.

Two Routes to Unevenness

The Big Bang that gave birth to the cosmos is presumed to have distributed matter evenly through space; scientists find evidence for this in the microwave radiation that comes with equal intensity from all directions. The puzzle is that astronomical surveys also reveal the universe to be very lumpy: Galaxies and clusters of galaxies seem to occur on the surfaces of interconnected, bubblelike voids, a configuration that some scientists have likened to Swiss cheese. Shown at right are two nearly opposite theories describing how cosmic strings could have produced this large-scale structure. One hypothesis *(top row)* holds that matter in the early, featureless universe congealed around the strings, drawn by their powerful gravity. The opposing theory *(bottom row)* posits that the pressure of the strings' electromagnetic radiation pushed matter away.

If strings were the underlying skeleton on which the universe was built, indirect evidence of their existence could come from observations of specific types of gravitational lensing *(below)*. Other proof, less easy to find, would be the whispering of gravity waves left behind by these now-vanished cosmic bones.

Astronomers looking for signs of cosmic strings search for double images produced by an effect known as gravitational lensing. The gravity of a massive object—a black hole, a supermassive galaxy, or a string—could bend light beams from more distant bodies such as quasars, projecting the light source as a double image. One indication that the lensing is the result of a string and not something else would be a row of twin images *(below)*, caused by a string lying between Earth and several quasars. With a black hole, for example, only one set of twins would appear.

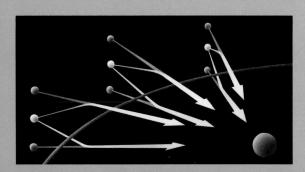

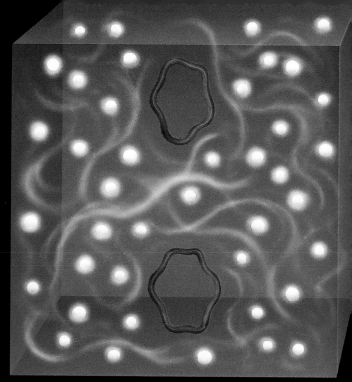

The attraction theory for strings shows the early universe as largely structureless; filled with uniformly distributed gas, stars, and star clusters; and containing massive but almost dimensionless flaws in the form of cosmic strings.

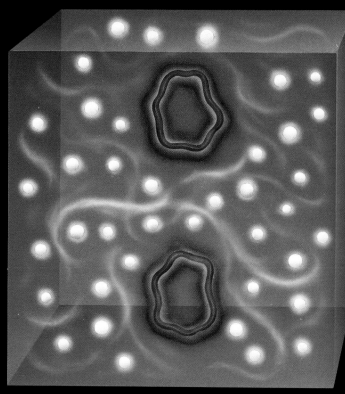

The repulsion theory for strings also describes matter in the early universe as being evenly distributed. But in this scenario, the strings' emission of electromagnetic radiation is more powerful than the attractive force of their massive ●. sity

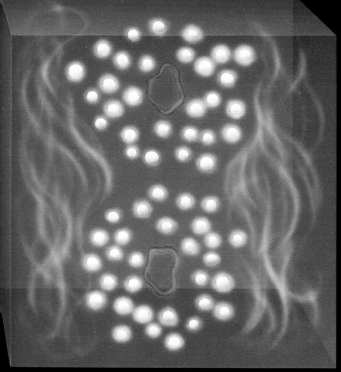

Attracted by the strings' gravity, gas and stars become concentrated around the loops, redistributing matter less evenly through the cosmos. As the strings oscillate, gravity waves carry away their energy, causing them to shrink.

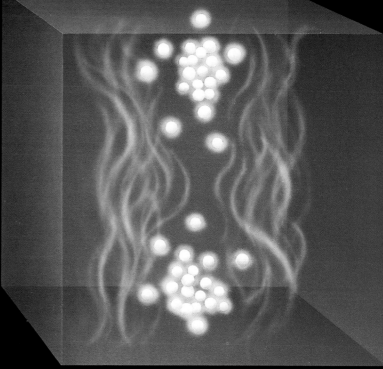

Continued dissipation of their store of energy causes the cosmic strings to evaporate out of existence, leaving voids surrounded by gas and star clusters that over billions of years will evolve into galaxies and galaxy clusters.

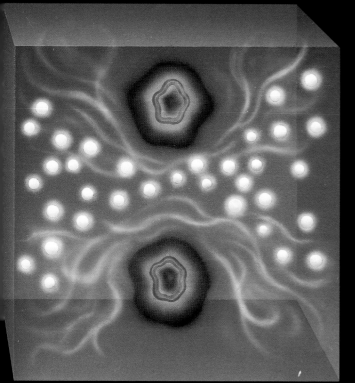

The radiation generated by the strings moves like a wave, pushing away surrounding gas and star clusters, sweeping out voids in space. Matter begins to pile up in the areas between voids, building density that sets the stage for galaxy formation.

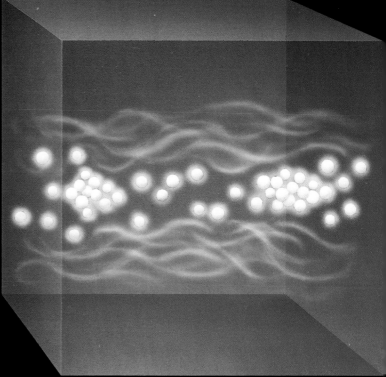

When the strings have dissipated all their energy and vanished, they leave behind a shell of star clusters and gas at the boundary where voids meet. The result is the same: Matter has been redistributed on a large scale, and the universe is lumpy.

cumulated data suggesting that the mass of the neutrino might be on the order of thirty electron volts.

Some fresh light was shed on the neutrino-mass question in 1987 by an exploding star, or supernova, 170,000 light-years from Earth. The theory of supernova dynamics holds that a giant star is rent by a catastrophic blast when its dense core collapses, producing an enormous amount of energy that is carried away by neutrinos. When supernova 1987A flared in the southern sky on February 23, scientists working with neutrino detectors immediately looked for signs of unusual activity. They were not disappointed. Supernova 1987A was so strong and so close that each square inch of the Earth's surface was showered with 100 billion neutrinos from the explosion. Even so, only a handful of those particles showed up in the detectors. In the underground salt mine near Cleveland, exactly eight neutrinos flashed in the proton decay detector; another eleven appeared in a similar device in Kamioka, Japan.

According to the theory of relativity, massless particles must all travel at light-speed. However, if a body has some mass, no matter how small, then its velocity must be less than the speed of light and proportional to its kinetic energy. In the case of the supernova neutrinos, this energy was imparted by the collapse of the star's core. Since it is unlikely that each particle received an identical jolt from the explosion, scientists expected the neutrinos to have different velocities. As a consequence, they would arrive on Earth at discernible intervals, which could be related to the variations in their energy levels. Since mass and velocity are the only components of kinetic energy, physicists would have new data for neutrino mass calculations.

As it happened, the data was meager: The nineteen neutrinos arrived in a span of just thirteen seconds. However, the scientific reaction was prodigious, with more than sixty papers appearing in a matter of months. Although they were based on disparate assumptions about the original conditions in the supernova and used a variety of mathematical techniques, most reached similar, necessarily inexact conclusions. The neutrinos from supernova 1987A might have mass, but it could be no greater than about twenty electron volts—barely enough to create Zel'dovich's pancakes. Nevertheless, the margin of error in the calculations left open the possibility that the neutrinos had no mass whatsoever.

EXPLAINING INVISIBLE MASS
Although the issue remains unresolved, the attribution of some mass to neutrinos might solve another nagging cosmological problem. Most astronomers believe that a substantial portion of the mass of the universe is made up of some undetermined kind of matter that cannot be observed by conventional means. Since neutrinos are effectively invisible and known to pervade the cosmos (estimates of their density range from one billion to ten billion per cubic meter), they would be good candidates for this so-called dark matter if they do have mass.

The first signs of the unseen matter had appeared in studies of celestial

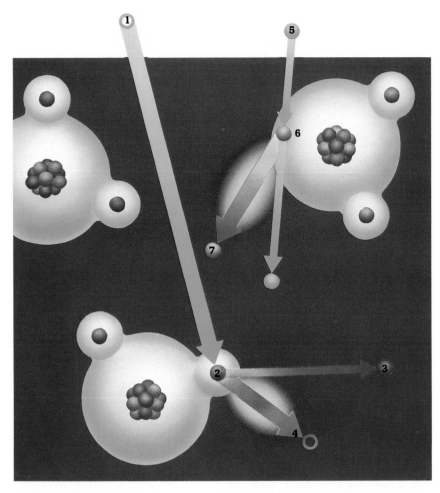

How to spot neutrinos. Exaggerated for clarity, the drawing above depicts a method of detecting neutrinos, tiny particles with little or no mass that bombard the Earth but rarely interact with any matter. Using large underground water tanks rigged with light sensors, physicists try to spot encounters between neutrinos and water molecules. In this example, an electron-antineutrino (1)—the opposite of an electron-neutrino—crashes into the single proton that makes up a hydrogen nucleus (2). The high-energy impact yields a neutron (3) and a particle called a positron, which emits a flash of radiation (4). At top right, an electron-neutrino (5) grazes an oxygen atom (6) and knocks loose an electron (7) with such force that, again, light is produced.

motions conducted during the 1930s. One, led by Dutch astronomer Jan Hendrik Oort, looked at the motion of stars in the outlying regions of the Milky Way. Oort calculated how much mass the inner part of the galaxy must have to keep those stars gravitationally bound in their orbits. He then estimated the actual mass of the innermost stars, coming up with a total 50 percent smaller than needed to account for the observed motions. Fritz Zwicky, a Swiss astronomer working during the same period in the United States, discovered similar aberrations on a scale 10,000 times larger. Measuring the positions and velocities of galaxies in large clusters, he found that the individual galaxies move so fast that their mutual gravitation is too weak to keep them from flying apart. Since there was no evidence that clusters were disintegrating, Zwicky concluded that they must include enough unseen mass to generate the attraction.

The question of the missing mass drifted unanswered in the field of observational astronomy until the 1980s, when new data began to accumulate. Vera Rubin of the Carnegie Institution in Washington, D.C., performed a careful analysis of the rotation of spiral galaxies, producing strong evidence that as much as 90 percent of their matter must be invisible. The debate over dark matter was revitalized, and this time cosmologists, by now deeply engaged in their own speculation about mass distribution, were drawn in.

The large amount of unseen mass must have played a significant role in the evolution of the universe. If that mass is made up of neutrinos, it can be explained within the framework of existing cosmological theory. But the standard model would have to be extensively revised if the anomalous motions are caused by ordinary, baryonic matter (so called because the particles that contribute most of its mass—protons and neutrons—belong to the family of quark-based particles collectively known as baryons). Extra baryons would invalidate an important part of the theory that accounts for the balance of elements detected in luminous matter. That balance has been shown to be a consequence of the synthesis of atomic nuclei in the early universe, in a sequence of events detailed by such theorists as George Gamow and Fred Hoyle.

Most astronomers therefore consider the dark matter to be nonbaryonic, ruling out such baryon-based candidates as black holes, gas clouds, and dim dwarf stars. Nonbaryonic alternatives to neutrinos have also been proposed.

One suggestion is a proliferation of cosmic strings, supposed relics of the Big Bang that retain that era's unfathomable density in one-dimensional domains *(page 112)*. Other possibilities are hypothetical offspring of advanced particle physics—strange creatures with names like axion, photino, and gravitino. Unlike neutrinos, though, neither the exotic particles nor the cosmic strings have ever appeared anywhere but in theories. Only new observations can decide the issue between neutrinos and their outlandish cousins.

TO THE END OF THE UNIVERSE

Regardless of its ingredients, the dark matter will be the most important player in the final act of the cosmic drama. It contributes to the gravitational pull of mass in the universe that has slowed the dilation of space-time since the Big Bang. If the invisible mass is sparse enough, the expansion will continue forever, winding down but never stopping. With passing eons, the sky will become darker as the stars in ever more distant galaxies wink out, their fuel exhausted. Then even the cinders will disappear, their atoms broken up one at a time by the slow decay of protons. In the immeasurably remote future, the cosmos will be utterly dark, cold, and empty.

An entirely different end awaits if the unseen matter makes the universe massive enough to collapse on itself. If the mass is twice the critical value, the expansion will coast to a halt in 50 billion years, having swelled the universe to twice its present size. As the contraction begins, the temperature of the background radiation will rise. When the universe reaches one-hundredth its present size, 59 billion years after the beginning of the contraction, the night sky will radiate as much energy as the Sun now pours onto the Earth at noon. Seventy million years later, the universe will be ten times smaller, and it will have become so bright and hot everywhere that gaseous molecules will begin to break up into their constituent atoms. These, in turn, will quickly be stripped of their electrons.

The heat will build at an accelerating pace. After another 700,000 years, it will reach 10 million degrees Kelvin, and even planets and stars will dissolve into the soup of radiation, electrons, and nuclei. Now hurtling in on itself, the universe will reach 10 billion degrees Kelvin in just twenty-two days. Atomic nuclei will begin to split into protons and neutrons, then into free quarks. In ever-dwindling intervals of time, the entire sequence of the original expansion will run in reverse. Finally, the whole cosmos will shrink into a single point, a chaotic domain of pure energy, and time itself will come to an end— or to the threshold of a new beginning.

Fiery or icy, the end of our universe, like its start, hovers at the far edge of human comprehension. The advance of science has removed these events from the province of mythological gods and giants: Now they are understood as extensions of the physical processes that govern all matter. Many questions remain, deep issues of how and why. But the progress made so far suggests that the better we are able to understand the cosmos, the more wondrous it will seem.

THE INFANT UNIVERSE

According to all the best reckoning, the universe began 15 to 20 billion years ago, when, say many theoreticians, space and time expanded from a single point in an indescribable burst of energy known as the Big Bang. Certain conditions of the Big Bang itself and the instant that followed it remain beyond the reach of contemporary science *(pages 104-107),* but cosmologists believe they can trace much of the story back to within an infinitesimal fraction of a second of the cosmic explosion. Using a mixture of astronomical observation, high-energy particle experiments, and theoretical physics, they have attempted to describe events starting at 10^{-43} second.

In the most generally accepted scenario, the rapid expansion of space immediately after the Big Bang caused the temperature of the universe to plummet from more than 10^{32} degrees Kelvin to only a billion degrees in about one minute. As temperatures fell past key values—analogous to freezing points—conditions snapped swiftly from one physical state to the next.

The pages that follow depict the evolution of the early universe through its initial, rapid-fire changes into the eras beyond. Within microseconds previously unified forces separated, new particles appeared, and old ones vanished. The pace of change then slowed for almost one million years, until the universe cooled enough to allow the creation of complete atoms. At this point, a major transition occurred: From an age when the energy of the universe was dominated by radiation, or light, the cosmos entered the age dominated by matter, and the foundations of today's universe were laid.

A Particle Primer

Events after the Big Bang involved a diverse host of elementary particles, represented by the simplified group at right. (A more detailed examination may be found on pages 83-89.)

Physicists divide particles into two categories: fermions, which typically carry matter, and bosons, which generally convey force. Fermions include quarks and leptons, and their antimatter counterparts (top three rows). Quarks are entities that combined into protons and neutrons. Leptons later evolved into distinct forms, including electrons and neutrinos.

In the earliest instants after the Big Bang, most forces were indistinguishable. As the forces separated, each acquired its own identity as a carrier boson (right). The strong nuclear force, which in effect glues quarks together, is conveyed by gluons; the weak force, responsible for radioactive decay, is transmitted by intermediate vector bosons. The electromagnetic force is carried by photons, while gravity, according to most physicists, may operate through elusive particles that are called gravitons.

The last two rows at right depict now-extinct bosons that were created in extreme conditions soon after the Big Bang. The leptoquark and the antileptoquark came and went during the GUT Era illustrated on these two pages. The X Higgs boson and the H Higgs boson played significant roles in the periods that came immediately after the GUT Era.

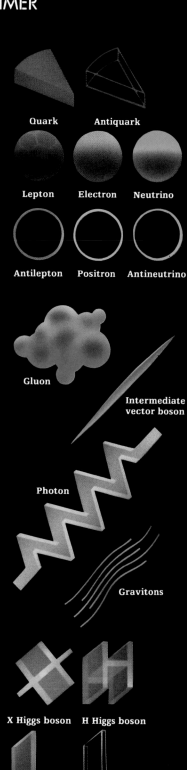

Quark Antiquark

Lepton Electron Neutrino

Antilepton Positron Antineutrino

Gluon

Intermediate vector boson

Photon

Gravitons

X Higgs boson H Higgs boson

Forces United

At 10^{-43} second after the Big Bang, the universe was a chaotic soup of energy-matter 10 trillion trillion times hotter than the core of an average star. In the next 10^{-35} second, particles of matter and their antimatter counterparts (left) sprang fleetingly into existence, only to vanish again in annihilating collisions that gave birth to yet more particles (opposite). Other encounters produced entities far more massive than any known today, including some that allowed particles to exchange their very identities.

This brief and energetic period is called the GUT Era, for several "grand unification theories" put forward by physicists who suggest that three of the four known forces—the electromagnetic and the strong and weak nuclear forces—were at the time still indistinguishable, or unified, in the electronuclear force. (The unification theories exclude gravity, which is thought to have assumed its separate identity just as the GUT Era began.) So dense was the cosmic broth at the end of the era that the mass of a cluster of galaxies would have fit easily into a volume smaller than that of a hydrogen atom.

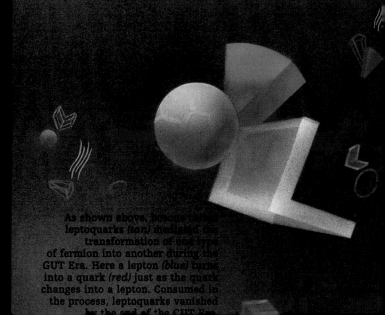

As shown above, bosons called leptoquarks (tan) mediated the transformation of one type of fermion into another during the GUT Era. Here a lepton (blue) turns into a quark (red) just as the quark changes into a lepton. Consumed in the process, leptoquarks vanished by the end of the GUT Era.

Collisions of enormous energy
during the GUT Era yielded a
shower of new particles. At left, for
example, a quark-lepton encounter produces a boson carrying the
unified electronuclear force
(marked here with the colors of its
three component forces). The crash
also creates a quark and antiquark,
and a leptoquark and antileptoquark, and—in a typically bizarre
quantum act—re-creates the original quark and lepton.

A lepton and antilepton annihilate
each other in a collision that could
have many possible outcomes;
two are shown here. One such encounter might produce a boson of
electronuclear force (above, right).
However, an identical collision
could yield fermions instead,
represented at right by a quark-
antiquark pair.

INFLATION'S END

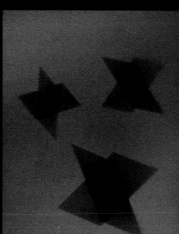

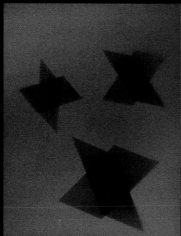

Trapped in a strange state known as a false vacuum, the universe seemed destined to expand forever at an exponentially increasing rate. That it did not is in part due to the tendency of all physical systems to seek the lowest available energy state. For the cosmos, this state is defined as one in which the electronuclear force is broken.

The transition coincided with the appearance of particles called X Higgs bosons *(left)*. The interaction between these bosons and the false vacuum led to a decline in the vacuum's latent energy and an increase in the mass of the particles.

The particles gained mass slowly at first *(top)*, and then more rapidly *(middle)*, until they materialized explosively out of the vacuum *(bottom)*. This reheated the universe to temperatures around those prevailing in the GUT Era and caused expansion to resume more normal rates.

During this change, some of the X Higgs bosons were absorbed by bosons of the unified electronuclear force *(right)*, yielding gluons and the electroweak force. This breakup of the electronuclear force compensated for the missed transition that had caused the false vacuum in the first place. Other X Higgs simply decayed in a particulate burst *(opposite)* as the universe moved out of the inflationary phase.

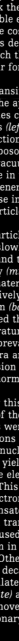

THE INFLATION ERA

The GUT Era ended at 10^{-35} second after the Big Bang as the cosmic temperature fell precipitously past a critical mark of about 10^{27} degrees Kelvin. So rapid was this supercooling that the electronuclear force, instead of breaking up, remained unified. The result was an unstable state known as a false vacuum. As the universe continued to expand, the temperature and energy of individual particles plummeted; paradoxically, however, the total energy of the universe grew. Combined with the increasing volume of space, this growth had a very peculiar effect on expansion. Instead of slowing, the rate of expansion skyrocketed. By the end of this so-called Inflation Era at 10^{-33} second after the Big Bang, the volume of space had increased more than a trillion trillion times. The details of how inflation ended are uncertain *(left)*, but eventually the unstable false vacuum gave way to the matter-dominated universe of today.

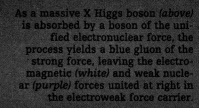

As a massive X Higgs boson *(above)* is absorbed by a boson of the unified electronuclear force, the process yields a blue gluon of the strong force, leaving the electromagnetic *(white)* and weak nuclear *(purple)* forces united at right in the electroweak force carrier.

In a collision typical of the highly energetic Inflation Era, a quark and lepton produce, in this instance, a gluon *(blue)*, gravitons *(light blue)*, an electroweak force carrier *(white and purple)*, a quark and antiquark *(both red)*, a lepton and antilepton *(both blue)*, and the original quark and lepton.

A weak Higgs boson decays into assorted particles, including quarks, leptons, antiquarks, and antileptons. As seen here, the process yielded more matter than antimatter, producing about a billion and one particles of matter for every billion of antimatter— a difference that had important consequences in the Quark Confinement Era (pages 126–127).

THE ELECTROWEAK ERA

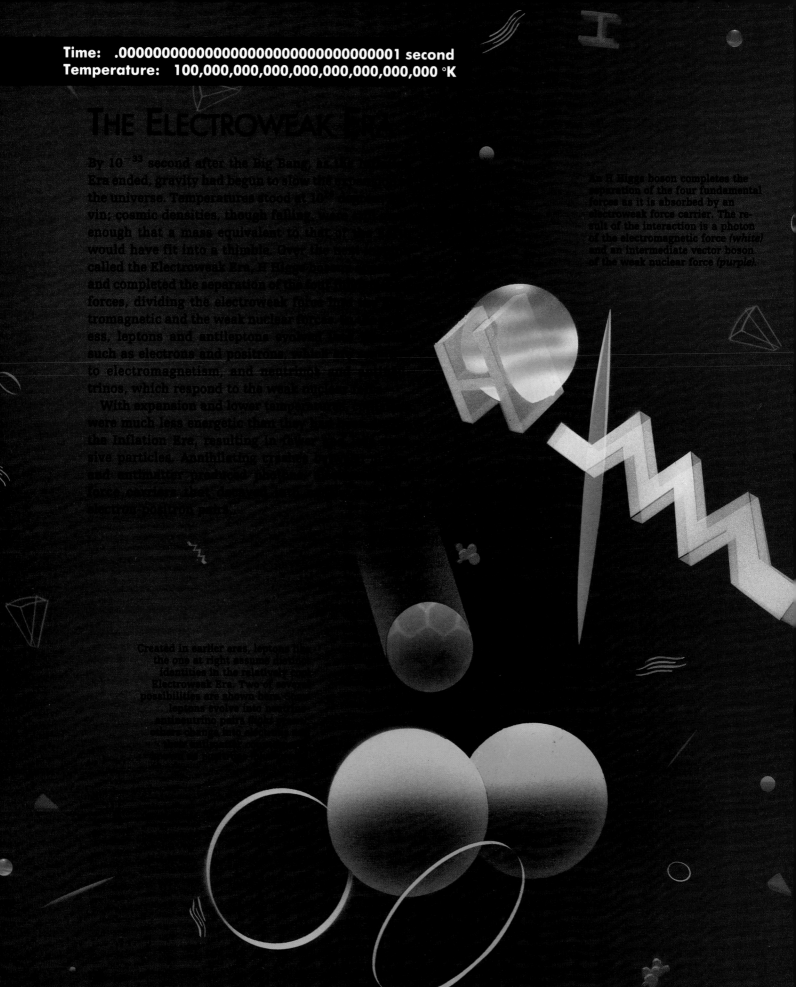

By 10^{-33} second after the Big Bang, the Inflation Era ended, gravity had begun to slow the expansion of the universe. Temperatures stood at 10^{27} degrees Kelvin; cosmic densities, though falling, were still high enough that a mass equivalent to that of the sun would have fit into a thimble. Over the period called the Electroweak Era, H Higgs bosons arose and completed the separation of the four fundamental forces, dividing the electroweak force into the electromagnetic and the weak nuclear forces. In the process, leptons and antileptons evolved into particles such as electrons and positrons, which respond to electromagnetism, and neutrinos and antineutrinos, which respond to the weak nuclear force.

With expansion and lower temperatures, collisions were much less energetic than they had been during the Inflation Era, resulting in fewer very massive particles. Annihilating clashes of matter and antimatter produced photons and other force carriers that decayed into new particles — electron-positron pairs.

An H Higgs boson completes the separation of the four fundamental forces as it is absorbed by an electroweak force carrier. The result of the interaction is a photon of the electromagnetic force (white) and an intermediate vector boson of the weak nuclear force (purple).

Created in earlier eras, leptons like the one at right assume distinctive identities in the relatively cool Electroweak Era. Two of several possibilities are shown here. Some leptons evolve into neutrino-antineutrino pairs (left); others change into electron and positron antimatter pairs.

... between an electron ... a quark during the Electro-weak Era produces a boson for each of the four forces. The collision also yields an electron and a positron, a quark and its antiquark, and the original quark and electron pair.

When an electron and a positron annihilate each other, the result is two high-energy photons *(white)*, each of which promptly decays into an identical electron-positron pair. Such processes continue as long as energy levels remain high, turning the cosmic energy into matter and antimatter.

CONFINING THE QUARKS

The basic matter of today's universe began to come together 10^{-6} second after the Big Bang as temperatures fell to about 10^{13} degrees Kelvin. Though still more than a million times hotter than the Sun's core, this relatively low energy level allowed gluons of the strong force to unite quarks into the building blocks of nuclei: protons, neutrons, and their antiparticles. The quarks remained imprisoned in nuclear particles by conditions that were cooler and less energetic than those in which they originated. Because neutrons occasionally decayed into protons, protons gradually came to outnumber neutrons. Matter and antimatter annihilations continued, but instead of producing more matter, many of these events produced photons too weak to yield new matter. However, these weak photons were still capable of blocking the formation of proton-electron bonds that would have led to formation of atoms. The slight excess of matter generated from the Inflation Era (pages 122-125) was critical to the future of the universe. When particle collisions produced annihilation, matter disappeared in ratios of the one-for-one...

In a typical transaction of the Quark Confinement Era, a colliding electron-positron pair (below) annihilate each other and release two high-energy photons. As in the Electroweak Era, the photons decay into two more electron-positron pairs—the last pairs to form from the process of particle annihilation.

...gluons bind quarks together (left) to form larger particles such as neutrons and protons. Generic quarks and gluons are shown here, but the actual process (pages 63-69) involves a variety of specific types of quarks and gluons.

At right, a newly formed proton and antiproton annihilate each other. In contrast to the product of an electron-positron collision (above), the resulting photons do not carry enough energy to reproduce relatively massive proton-antiproton pairs. Because no new antimatter is created, the small excess of matter generated in the Inflation Era becomes dominant.

Despite their electrical attraction, a positively charged proton and negatively charged electron are unable to bond, stopped by one of the photons created in the Quark Confinement Era.

THE NEUTRINO ERA

For fifty-eight seconds after the confinement of the quarks *(left)*, the universe entered what might be called the Neutrino Era. The creation of electrons and positrons ceased for lack of energy, and as positrons, like other antimatter, gradually disappeared, the only antiparticle left was the antineutrino.

Both neutrinos and antineutrinos, which evolved during the Electroweak Era, stopped interacting with other particles of matter and thus became almost impossible to observe. Chargeless and perhaps massless, they continue today to pass through space, Earth, and even human bodies in virtually undetectable hordes that are presumed to travel at the speed of light.

Time: 2 seconds to 1 minute
Temperature: 10,000,000,000 °K to 1,300,000,000 °K

Survivors from the first minute, neutrinos and antineutrinos are impervious to most physical influences. Subject only to the weak force and to the feeble tug of gravity, they pass through matter as if it did not exist.

Forging Atomic Nuclei

In the Nucleosynthesis Era, which began one minute after the Big Bang and lasted about four minutes, conditions finally were ripe for the creation of the first atomic nuclei. At the three-minute mark, the density of the universe resembled that of water, and by the end of the era temperatures had fallen to 600 million degrees Kelvin. In the era's most critical development, photons began to lose more of their energy; thus depleted, they could no longer prevent protons and neutrons from combining into atomic nuclei. Even in this less energized state, however, photons retained sufficient power—given continued expansion and cooling—to prevent nuclei from combining with electrons to form whole atoms. As protons and neutrons came together, traces of other elements emerged, but the most common groupings were varieties of hydrogen and helium, which account for most of the known matter in the cosmos today. (No heavier elements were formed because the expanding universe cooled too quickly to allow further nuclear fusion.)

2. A secondary nucleosynthesis occurs as a free proton encounters a deuteron, binding with it into a helium-3 nucleus. Most of the helium nuclei presently in existence probably formed at this time.

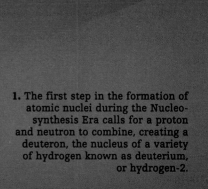

1. The first step in the formation of atomic nuclei during the Nucleosynthesis Era calls for a proton and neutron to combine, creating a deuteron, the nucleus of a variety of hydrogen known as deuterium, or hydrogen-2.

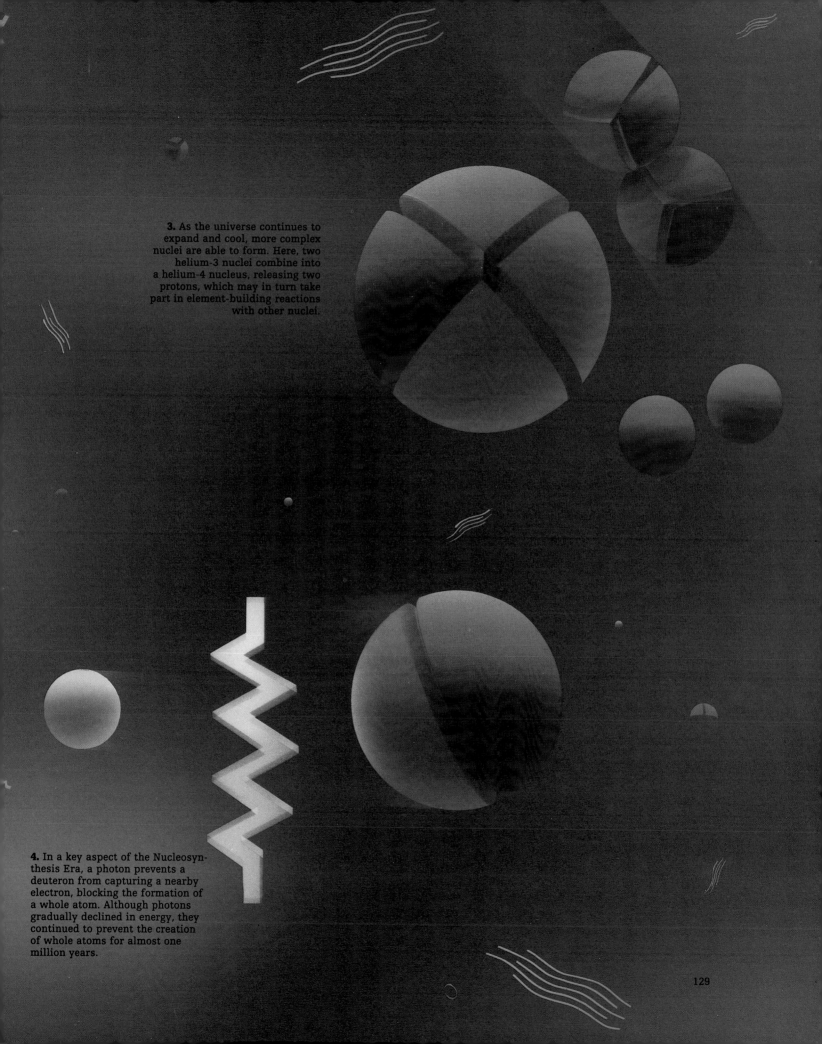

3. As the universe continues to expand and cool, more complex nuclei are able to form. Here, two helium-3 nuclei combine into a helium-4 nucleus, releasing two protons, which may in turn take part in element-building reactions with other nuclei.

4. In a key aspect of the Nucleosynthesis Era, a photon prevents a deuteron from capturing a nearby electron, blocking the formation of a whole atom. Although photons gradually declined in energy, they continued to prevent the creation of whole atoms for almost one million years.

THE AGE OF MATTER

By the end of the Nucleosynthesis Era, five minutes after the Big Bang, the pace of change had slowed dramatically. The universe continued to expand and cool, but no significant transitions would occur for about one million years. The cosmic density was approximately that of air, and temperatures had dropped from 10^8 degrees to just 3,000 degrees Kelvin. At that point, drastically weakened photons could no longer disrupt the formation of atoms. Positively charged nuclei and negatively charged electrons were at last able to unite into atoms, giving rise to the present Matter Era.

One result of atom formation was the gradual clearing of the cosmic plasma fog. As free electrons attached themselves to nuclei, photons were no longer scattered by random encounters with electrons, and space became transparent. The energies of photons continued to decline, dropping over the next 15 to 20 billion years to the three degrees Kelvin radiation that permeates the universe today.

THE FIRST ATOMS

During the Matter Era, nuclei and electrons united to make atoms. The particular combination of protons and neutrons in a given nucleus determined the kind of atom. Since the energetic conditions just after the Big Bang allowed the formation of only a few combinations, atomic varieties were limited to the assortment below, ranked in order of increasing mass. Atoms of all other elements would not be fused until many eons later, in the cores of stars.

With the continued expansion of space, photons gradually lose their energy and with it their ability to keep nuclei and electrons apart. As illustrated in the sequence below, the mutual attraction between a tiny electron and a larger nucleus overwhelms a depleted photon—and an atom is born.

A single-proton hydrogen nucleus captures one electron to produce a hydrogen atom, the simplest possible.

A deuteron of one proton and one neutron attracts an electron, creating an atom of deuterium, or heavy hydrogen.

With the addition of an electron, a nucleus made up of two neutrons and one proton forms a stable atom of tritium.

Two electrons balance the charge of two protons in this single-neutron nucleus, forming an atom of helium-3.

The most common form of helium is formed by the combination of two protons, two neutrons, and two electrons.

Adding three electrons to a three-proton, two-neutron nucleus creates lithium, the most massive atom of the era.

GLOSSARY

Acceleration: a change in velocity. The term includes changes of direction and decreases as well as increases in speed.

Alpha particle: the nucleus of one form of helium, consisting of two protons and two neutrons.

Annihilation: the disappearance of a particle and matching antiparticle as a result of their collision. The collision converts the masses of the particle and antiparticle into energy.

Antilepton: an antiparticle to any of several types of leptons.

Antileptoquark: in theory, a boson that is the antiparticle to the leptoquark. Like its counterpart, it would have existed only for an instant in the early universe.

Antimatter: matter made up of antiparticles. Most antimatter would have been annihilated in the first second after the Big Bang.

Antiparticle: a particle identical in mass to a matter particle but opposite to it in properties such as electrical charge. For example, a positron is the antiparticle to an electron.

Antiproton: an antiparticle to a proton.

Antiquark: an antiparticle to a quark. Antiquarks have anticolor instead of color.

Arc second: a sixtieth of an arc minute, which is in turn a sixtieth of a degree of arc; there are 360 degrees in a circle. Arc seconds, minutes, and degrees measure an object's apparent size and position on the sky.

Atom: the smallest component of a chemical element that retains the properties associated with that element. Atoms are composed of protons, neutrons, and electrons; the number of protons determines the identity of the element.

Atomic weight: the mass of an atom, roughly equal to the number of protons and neutrons in its nucleus.

Background limit: the distance beyond which objects form a continuous screen; if objects do not extend to the limit, gaps will be observed between them. In an infinite universe, stars would extend past the background limit, filling the night sky with continuous starlight.

Background radiation: a detectable, steady emission of electromagnetic radiation from all directions of the sky. Cosmic background radiation in microwave frequencies is commonly attributed to the Big Bang.

Big Bang: according to a widely accepted theory, the primeval moment, 15 to 20 billion years ago, when the universe began expanding from a state of infinite density.

Black body: a hypothetical object that absorbs and reemits all radiation reaching it.

Black hole: in theory, an extremely compact body with such great gravitational force that no radiation can escape from it.

Boson: a particle that carries force; one of two types of elementary particles.

Bubble chamber: an experimental device that maintains a quantity of liquid just at the point of boiling, so that passing high-energy particles will leave trails of bubbles.

Celsius: a scientific temperature scale in which 0 degrees is the freezing point and 100 degrees the boiling point of water.

Closed universe: a universe in which the average density of mass exceeds a critical value, so that gravity will eventually reverse the expansion of space. According to relativity theory, the geometry of such a universe is spherical.

Cluster (of galaxies): a gravitationally bound system of galaxies, ranging in number from a few dozen to several thousand.

Color: an abstract property, not related to ordinary visible color, that governs the interaction of some elementary particles. Particle color may be assigned the value blue, red, or green.

Cosmic singularity: in theory, the state of the universe before the Big Bang, when all matter was compressed into a state of infinite density.

Cosmic string: according to theory, a type of massive, one-dimensional object that formed during the early expansion of the universe.

Cosmological constant: a mathematical factor introduced by Einstein into the field equations of general relativity to accommodate his belief in a static universe. Today the cosmological constant is predicted by some theories but is normally set at zero and thus disregarded.

Cosmological principle: the assumption that the universe is homogeneous at the very largest scale, appearing essentially the same to observers in all locations.

Cosmological red shift: *see* Red shift.

Cosmology: the study of the universe as a whole, including its large-scale structure and movements, origin, evolution, and ultimate fate. A specialist in this field is called a cosmologist.

Cosmos: the universe; also, a mathematical or scientific model of the universe.

Critical density: a crucial value for the density of matter in the universe, about 4.5 times 10^{-29} gram per cubic centimeter. The relation between the actual mass density of the universe and the critical density determines whether the universe is closed, flat, or open. *See* Closed universe, Flat universe, Open universe.

Curved space: a distortion in the geometry of space, normally occurring near massive bodies, that can be detected by observing the curved path followed by light in that region.

Dark matter: a form of matter that has not yet been directly observed but whose existence is deduced from its gravitational effects.

Decay: the spontaneous transformation of a particle into one or more other particles, which may then decay as well.

Density: the amount of matter in a given volume of space.

Deuteron: the nucleus of an atom of deuterium, an isotope of hydrogen. A deuteron contains one proton and one neutron.

Doppler effect: a phenomenon in which waves appear to compress as their source approaches the observer or stretch out as the source recedes from the observer. *See* Red shift.

Electromagnetic radiation: radiation consisting of periodically varying electric and magnetic fields that vibrate perpendicularly to each other and travel through space at the speed of light.

Electromagnetic spectrum: the array of electromagnetic radiation, ranging in order of frequency or wavelength from low-frequency, long-wavelength radio waves through infrared radiation, visible light, and ultraviolet radiation to high-frequency, short-wavelength gamma rays.

Electromagnetism: the force that attracts oppositely charged particles and repels similarly charged particles. Electromagnetism affects all charged particles but not neutral particles such as neutrinos.

Electron: a negatively charged particle that normally orbits an atom's nucleus but may exist in isolation.

Electronuclear force: according to grand unification theories (GUTs), a force in which electromagnetism, the weak force, and the strong force are combined and indistinguishable. The electronuclear force could exist only at the extremely high energies of the very early universe or in particle accelerators.

Electroweak force: a force in which electromagnetism and the weak force are combined and indistinguishable. The electroweak force can exist only at very high energies.

Element: one of just over 100 substances that cannot be reduced by chemical means to simpler substances.

Elementary particle: a fundamental, irreducible component of the physical universe. The two currently recognized classes of elementary particles are bosons and fermions.

Ellipse: a closed, symmetrical curve with two focal points and with vertical and horizontal axes of unequal length.

Energy: the ability to do work, where work is defined as moving mass through space.

Energy level: a discrete quantity of energy associated with a particle within an atom or a nucleus. An increase in energy will shift electrons to higher energy levels within the atom.

Equivalence principle: the rule, derived from general relativity, that in a small region of space-time, the effects of a gravitational field are indistinguishable from those produced by an acceleration of the frame of reference.

Ether: a hypothetical substance capable of carrying light waves, once thought to permeate all of space.

Euclidean geometry: *see* Geometry.

Fahrenheit: a nonastronomical temperature scale in which 32 degrees is equivalent to 0 degrees Celsius and 212 degrees is equivalent to 100 degrees Celsius.

Fermion: a particle of matter or antimatter; one of two classes of elementary particles.

Field: the influence exerted by a force, such as electromagnetism or gravity, throughout a region of space. A field has a precise value at all points in space-time.

Field equation: one of the complex equations used to describe the space-time contours of gravitational and other force fields.

Flat universe: a universe in which the average density of mass matches the critical density, so that gravity is just barely too weak to stop the universe's expansion. According to relativity theory, the geometry of such a universe is flat. *See* Critical density.

Force: a physical phenomenon that can change the momentum of an object. The four accepted present-day forces are gravity, electromagnetism, the strong force, and the weak force.

Frame of reference: a position from which objects are viewed.

Frequency: the number of oscillations per second of an electromagnetic (or other) wave. *See* Wavelength.

Galaxy: a system that contains stars numbering from millions to hundreds of billions as well as varying quantities of gas and dust.

General relativity: a theoretical account of the effects of acceleration and gravity on the motion of bodies and the observed structure of space and time.

Geodesic: the shortest path between two points. On a flat surface, a geodesic is a straight line; on a spherical surface, an arc.

Geometry: a set of rules describing the structure of space in a given region. Classical *Euclidean geometry* applies to a flat surface or to "flat" space; *non-Euclidean geometries* apply to curved surfaces or curved space and may include such unlikely phenomena as triangles whose vertices total less than 180 degrees.

Gluon: a boson, or force-carrying particle, that conveys the strong force and binds quarks together. There are eight types of gluons.

Grand unification theory (GUT): any of several competing but similar theories that unite electromagnetism, the weak force, and the strong force into one electronuclear force.

Gravitational lens: an optical effect of general relativity in which the gravity of a very massive body bends the light of an object behind it, distorting its apparent image and often producing one or more duplicate images.

Gravitational red shift: *see* Red shift.

Graviton: according to theory, a boson that transmits the force of gravity. Gravitons have not been experimentally verified.

Gravity: the force responsible for the mutual attraction of separate masses.

Gravity wave: a theoretical perturbation in an object's gravitational field that would travel at the speed of light. General relativity predicts that gravity waves may result from accelerating, oscillating, or violently disturbed masses, categories that include black holes and cosmic strings.

Gravity well: a local distortion in the fabric of space-time near a massive body, analogous to a well or depression in a two-dimensional sheet.

H Higgs boson: in theory, a massive boson capable of transforming the electroweak force into distinct electromagnetic and weak forces. H Higgs bosons would have existed only from 10^{-33} to 10^{-12} second after the Big Bang.

Heisenberg uncertainty principle: *see* Uncertainty principle.

Helium: the second lightest chemical element and the second most abundant, with a nucleus that includes two protons and at least two neutrons.

Hydrogen: the most common detectable element in the universe, with a nucleus that includes one proton. In theory, hydrogen was the primary element produced just after the Big Bang.

Inertia: an object's tendency to stay in motion, if in motion, or at rest, if at rest; its resistance to acceleration.

Inflation: according to theory, a sudden expansion in space that occurred 10^{-35} second after the Big Bang.

Intermediate vector boson: a boson, or force carrier, that conveys the weak force. Charged intermediate vector bosons are also called W particles; neutral intermediate vector bosons are Z particles.

Inverse-square law: the mathematical description of how the strength of some forces, including electromagnetism and gravity, changes in inverse proportion to the square of the distance from the source.

Ion: an atom that has lost or gained one or more electrons. In comparison, the neutral atom has an equal number of electrons and protons, giving it a zero net electrical charge. A positive ion has fewer electrons than the neutral atom; a negative ion has more.

Isotope: one of two or more forms of a chemical element that have the same number of protons but a different number of neutrons in the nucleus.

Kelvin: an absolute temperature scale that uses Celsius degrees but sets 0 degrees at absolute zero, about minus 273 degrees Celsius.

Kinetic energy: an object's energy of motion.

Length contraction: the tendency of a moving object to shorten in the direction of its motion, as viewed by an observer who is stationary relative to the object.

Lepton: a fermion that is unaffected by the strong force. Leptons include electrons, muons, taus, and neutrinos.

Leptoquark: according to theory, a massive boson that enabled leptons and quarks to exchange identities. Leptoquarks would have occurred only during a brief period that began 10^{-43} second after the Big Bang.

Light: the visible part of the electromagnetic spectrum; may also refer to the entire electromagnetic spectrum, as in the "speed of light," the speed of all forms of electromagnetic radiation.

Light-year: an astronomical distance unit equal to the distance light travels in a vacuum in one year, almost six trillion miles.

Lorentz transformations: equations that relate the measures of length, time, and mass from one uniformly moving frame of reference to another.

Mass: a measure of the total amount of material in an object, determined either by its gravity or by its tendency to resist acceleration.

Mass increase: the tendency of a moving object to become more massive, as perceived by an observer who is stationary relative to the object.

Matter: the category of all fermion particles, as opposed to antiparticles; may also refer generically to both matter and antimatter.

Molecule: the smallest unit of a compound that retains the properties of that substance. A molecule may consist of a single atom or of two or more atoms bonded together.

Momentum: a measure of an object's inertia; an object's mass multiplied by its velocity.

Muon: a charged lepton analogous to an electron but much less stable.

Neutrino: one of a class of neutral leptons with little or no mass.

Neutron: a neutral particle, made up of three quarks, with a mass similar to that of a proton; normally found in an atom's nucleus.

Non-Euclidean geometry: see Geometry.

Nucleosynthesis: the combining of protons and neutrons to form the atomic nuclei of chemical elements.

Nucleus: the massive center of an atom, composed of protons and neutrons and orbited by electrons.

Observation limit: the greatest distance from which light now being received on Earth could have originated.

Open universe: a universe that is on average less dense than the critical density and so continues to expand. According to relativity theory, the geometry of such a universe is saddle-shaped. *See* Critical density.

Orbit: the path of an object revolving around another object.

Particle: a fundamental component of matter, antimatter, or force, such as a proton, neutron, lepton, or boson.

Particle accelerator: a device, often several miles long, used to accelerate subatomic particles to high velocities and fire them at other particles or at targets. The results of collisions suggest the particles' properties.

Particle physics: the experimental study of subatomic particles, often using particle accelerators.

Perihelion: the point in its orbit where a planet is closest to the Sun.

Phase transition: a complete change from one physical state to another. Freezing is a phase transition from a liquid to a solid state.

Photoelectric effect: the fact that light at certain frequencies can knock electrons out of a charged metal sheet.

Photon: a force carrier, or boson, that conveys electromagnetic force, or radiation, and is associated with a specific frequency.

Planck Era: theoretically, the very brief period after the Big Bang and up to the Planck Time. Conditions during the Planck Era violate the rules of conventional quantum mechanics and general relativity and cannot be adequately described by current physics.

Planck's constant: a number whose value is important to the equations of quantum mechanics; equal to the ratio of a photon's energy to its frequency.

Planck Time: in theory, an instant 10^{-43} second after the Big Bang, after which the universe would have followed conventional physical laws.

Positron: an antiparticle to the electron, carrying a positive electric charge.

Proton: a positively charged particle made up of three quarks, with about 2,000 times the mass of an electron; normally found in an atom's nucleus.

Quantum: a fixed packet, or quantity, of some physical property such as mass or energy.

Quantum chromodynamics (QCD): a theory that explains the strong force interactions between quarks in quantum terms.

Quantum electrodynamics (QED): a theory that explains electromagnetic interactions between particles in quantum terms.

Quantum gravity: a still-unrealized explanation of gravity in quantum mechanical terms, including its transmission by discrete particles called gravitons. Quantum gravity is crucial to the study of the Planck Era of the early universe.

Quantum mechanics: a mathematical description of the rules by which subatomic particles interact, decay, and form atomic or nuclear objects.

Quark: a fermion that responds to the strong force and so is never found in isolation. The six quark varieties are up, down, charm, strange, top, and bottom.

Radio: the least energetic form of electromagnetic radiation, having the lowest frequency and the longest wavelength.

Radio astronomy: the observation and study of radio waves produced by astronomical phenomena.

Red shift: a stretching of the apparent wavelength of light, which shifts its spectral lines toward the red end of the spectrum. A *Doppler red shift* is caused by the motion of a light source; a *cosmological red shift,* by the expansion of space between the observer and the light source; and a *gravitational red shift,* by the time-distorting effects of the gravity of massive bodies.

Relativistic velocity: motion at a significant fraction of the speed of light, in which changes in time, length, and mass become noticeable to observers who are stationary in relative terms.

Relativity: a set of theories that show how measurements are affected by motion and gravity. *See* General relativity, Special relativity.

Renormalization: a mathematical procedure in quantum physics that redefines the mass and charge of elementary particles in order to avoid certain "infinite" predictions.

Solar eclipse: the obscuring of the Sun's disk as the Moon passes directly between the Earth and the Sun.

Space-time continuum: a four-dimensional system that incorporates three spatial dimensions plus time.

Space-time event: an object's spatial position at an assigned instant in time.

Space-time interval: a mathematical expression of the distance in space and time between two points (events) on a worldline.

Special relativity: a theory postulating that observers in uniform motion cannot perceive their motion and that all observers in such motion obtain the same value for the speed of light. From these two principles the theory concludes that measures of distance, time, and mass will vary depending on the motion of an observer moving uniformly in relation to the thing being measured.

Spectral line: a bright or dark band in an astronomical spectrum, produced by atoms as they emit or absorb light.

Spectroscopy: the study of spectra, including the position and intensity of spectral lines.

Spectrum: the array of colors or frequencies obtained by dispersing light from a star or other source, as through a prism; often banded with spectral lines.

Spin: an abstract property of subatomic particles analogous to the angular momentum of a spinning top. Spin can be fractional, and it may be positive or negative.

Steady State model: an alternate theory to the Big Bang, not widely accepted, stipulating that the universe has always existed, and will always exist, in a state similar to the present.

Strong force: the force that binds quarks together into composite particles and holds protons and neutrons together to form atomic nuclei.

Subatomic particle: any particle smaller than an atom, from atomic components such as protons to the constituents of protons, quarks.

Supercluster: a large association of clusters of galaxies.

Supercooling: a phenomenon in which a substance cools so rapidly that there is not enough time for a phase transition such as freezing to occur at the expected temperature.

Superstring: a hypothetical constituent of fundamental particles in the form of a closed loop of energy. Superstrings would vibrate and manifest properties in many dimensions of space.

Supersymmetry: a theory in particle physics proposing that every type of fermion or boson has a matching partner particle differing from it only in spin.

Tau: a negatively charged lepton, similar to an electron or a muon but much more massive and very short-lived.

Thought experiment: a hypothetical series of simplified events, often impractical to replicate physically, in which logic is used to predict the outcome.

Time dilation: a relativistic effect in which the passage of time in a moving body slows from the viewpoint of a stationary observer.

Tritium: an isotope of hydrogen, with a nucleus containing one proton and two neutrons.

Ultraviolet: a band of electromagnetic radiation with a higher frequency and shorter wavelength than visible blue light. Most ultraviolet is absorbed by the Earth's atmosphere, so ultraviolet astronomy is normally performed in space.

Uncertainty principle: the fact that uncertain values are inevitable at the subatomic level since measuring techniques disrupt the particles being measured. For example, a particle's precise position and velocity can never be known at the same time.

Unified field theory: a theory that would explain all the field forces as different manifestations of a single, universal force. Its development is an unrealized goal of theoretical physicists.

Uniform motion: motion at a constant speed in a straight line without rotation. Uniform motion is the only kind of motion addressed by special relativity.

Universe: the totality of all matter and radiation and the space it occupies. The universe is believed to have a finite age of 15 to 20 billion years.

Vacuum: the lowest possible energy state.

Velocity: the speed and direction of motion.

Virtual particle: an extremely short-lived particle created out of nothingness, as permitted by the uncertainty principle. Although it exists too briefly to be directly observed, the effects of its existence may be detected.

Wave: the propagation of a pattern of disturbance.

Wavelength: the distance from crest to crest or trough to trough of an electromagnetic or other wave. Wavelengths are related to frequency: the longer the wavelength, the lower the frequency.

Weak force: a very short-range force responsible for particle decay.

Worldline: the path through space and time of a space-time event.

Wormhole: a hypothetical distortion in space-time linking widely separated black holes.

W particle: *see* Intermediate vector boson.

X Higgs boson: theoretically, a massive boson capable of releasing the strong force from the previously unified electronuclear force. X Higgs bosons would have existed only during a brief period beginning 10^{-35} second after the Big Bang; their decay would have produced an excess of matter over antimatter particles that is still evident.

X-rays: a band of electromagnetic radiation intermediate in wavelength between ultraviolet radiation and gamma rays. Because x-rays are completely absorbed by the atmosphere, x-ray astronomy must be performed in space.

Z particle: *see* Intermediate vector boson.

BIBLIOGRAPHY

Books

Abell, George O., David Morrison, and Sidney C. Wolff, *Exploration of the Universe.* Philadelphia: Saunders College Publishing, 1987.

Barnett, Lincoln, *The Universe and Dr. Einstein.* New York: Time, 1948.

Barrow, John D., and Joseph Silk, *The Left Hand of Creation.* New York: Basic Books, 1983.

Bartusiak, Marcia, *Thursday's Universe.* New York: Times Books, 1986.

Bernstein, Jeremy:
Einstein. New York: Penguin Books, 1973.
Three Degrees above Zero. New York: Charles Scribner's Sons, 1984.

Bondi, Harmann, et al., *Rival Theories of Cosmology.* London: Oxford University Press, 1960.

Bonnor, William B., *The Mystery of the Expanding Universe.* New York: Macmillan, 1964.

Bronowski, J., *The Ascent of Man.* Boston: Little, Brown, 1973.

Chandrasekhar, S., *Eddington.* Cambridge: Cambridge University Press, 1983.

Close, Frank, Michael Marten, and Christine Sutton, *The Particle Explosion.* New York: Oxford University Press, 1987.

Cosmology + 1: Readings from Scientific American. San Francisco: W. H. Freeman, 1977.

Crease, Robert P., and Charles C. Mann, *The Second Creation.* New York: Macmillan, 1986.

Crowther, James Gerald, *British Scientists of the Twentieth Century.* London: Routledge & Kegan Paul, 1952.

Douglas, A. Vibert, *The Life of Arthur Stanley Eddington.* London: Thomas Nelson and Sons, 1956.

Dukas, Helen, and Banesh Hoffmann, eds., *Albert Einstein: The Human Side.* Princeton, N.J.: Princeton University Press, 1979.

Eddington, Arthur, *The Expanding Universe.* Ann Arbor: University of Michigan Press, 1958.

Feinberg, Gerald, *What Is the World Made Of?* New York: Anchor Press/Doubleday, 1978.

Ferris, Timothy, *The Red Limit.* New York: Quill, 1983.

Feynman, Richard P., *"Surely You're Joking, Mr. Feynman!" Adventures of a Curious Character.* Toronto: Bantam Books, 1985.

Fritzsch, Harald, *The Creation of Matter.* New York: Basic Books, 1984.

Gamow, George, *The Creation of the Universe.* New York: New American Library, 1952.

Gillispie, Charles Coulston, ed., *Dictionary of Scientific Biography.* Vols. 3, 5, 7, 10, and 11. New York: Charles Scribner's Sons, 1981.

Godart, O., and M. Heller, *Cosmology of Lemaître.* Tucson, Ariz.: Pachart, 1985.

Goldberg, Stanley, *Understanding Relativity.* Boston: Birkhäuser, 1984.

Gravitation, Cosmology, and Cosmic-Ray Physics. Washington, D.C.: National Academy Press, 1986.

Gribbin, John, *In Search of the Big Bang.* Toronto: Bantam Books, 1986.

Guth, Alan H., "The Birth of the Cosmos." In *Origins and Extinctions,* ed. by Donald E. Osterbrock and Peter H. Raven. New Haven, Conn.: Yale University Press, in press.

Guth, Alan H., and Paul J. Steinhardt, "The Inflationary Universe." In *The New Physics,* ed. by Paul Davies. New York: Cambridge University Press, in press.

Harrison, Edward R.:
Cosmology. Cambridge: Cambridge University Press, 1982.
Darkness at Night. Cambridge, Mass.: Harvard University Press, 1987.

Hartmann, William K., *Astronomy: The Cosmic Journey.* Belmont, Calif.: Wadsworth, 1987.

Hawking, Stephen W., *A Brief History of Time.* Toronto: Bantam Books, 1988.

Henbest, Nigel, ed., *Observing the Universe.* Oxford: Basil Blackwell & New Scientist, 1984.

Hey, Anthony J. G., *The Quantum Universe.* Cambridge: Cambridge University Press, 1987.

Hoffmann, Banesh, *Relativity and Its Roots.* New York: Scientific American Books, 1983.

Hoffmann, Banesh, and Helen Dukas, *Albert Einstein: Creator and Rebel.* New York: Viking Press, 1972.

Hoyle, Fred, *The Nature of the Universe.* New York: Harper, 1950.

Islam, Jamal N., *The Ultimate Fate of the Universe.* Cambridge: Cambridge University Press, 1983.

Kaufmann, William J., III:
Relativity and Cosmology. New York: Harper & Row, 1973.
Universe. New York: W. H. Freeman, 1987.

Kevles, Daniel J., *The Physicists.* New York: Alfred A. Knopf, 1978.

Kolb, Edward W., et al., *Inner Space/Outer Space.* Chicago: University of Chicago Press, 1986.

Kutter, G. Siegfried, *The Universe and Life.* Boston: Jones and Bartlett, 1987.

Lang, Kenneth R., and Owen Gingerich, eds., *A Source Book in Astronomy and Astrophysics, 1900-1975.* Cambridge, Mass.: Harvard University Press, 1979.

Layzer, David, *Constructing the Universe.* New York: Scientific American Books, 1984.

Lemaître, Georges, *The Primeval Atom.* New York: D. Van Nostrand, 1950.

Mauldin, John H., *Particles in Nature: The Chronological Discovery of the New Physics.* Blue Ridge Summit, Pa.: Tab Books, 1986.

Merleau-Ponty, Jacques, and Bruno Morando, *The Rebirth of Cosmology.* New York: Alfred A. Knopf, 1976.

Mook, Delo E., and Thomas Vargish, *Inside Relativity.* Princeton, N.J.: Princeton University Press, 1987.

Motz, Lloyd, and Anneta Duveen, *Essentials of Astronomy.* New York: Columbia University Press, 1977.

Nicolson, Iain, *Gravity, Black Holes and the Universe.* New York: John Wiley & Sons, 1981.

Novikov, I. D., *Evolution of the Universe.* Cambridge: Cambridge University Press, 1983.

Pagels, Heinz R.:
The Cosmic Code. New York: Simon and Schuster, 1982.
Perfect Symmetry: The Search for the Beginning of Time. New York: Simon and Schuster, 1985.

Pais, Abraham, *'Subtle Is the Lord . . .': The Science and the Life of Albert Einstein.* Oxford: Oxford University Press, 1982.

Reines, Frederick, ed., *Cosmology, Fusion & Other Matters.* Boulder: Colorado Associated University

Press, 1972.

Schilpp, Paul Arthur, ed. and transl., *Albert Einstein: Autobiographical Notes.* La Salle, Ill.: Open Court, 1979.

Schwartz, Joe, *Einstein for Beginners.* New York: Pantheon Books, 1979.

Schwinger, Julian, *Einstein's Legacy.* New York: Scientific American Books, 1986.

Shapley, Harlow, Samuel Rapport, and Helen Wright, eds., *A Treasury of Science.* New York: Harper & Brothers, 1958.

Silk, Joseph, *The Big Bang.* San Francisco: W. H. Freeman, 1980.

Silverberg, Robert, *Niels Bohr: The Man Who Mapped the Atom.* Philadelphia: Macrae Smith, 1965.

Snow, C. P., *The Physicists.* Boston: Little, Brown, 1981.

Snow, Theodore P., *Essentials of the Dynamic Universe.* St. Paul, Minn.: West, 1987.

Terzian, Yervant, and Elizabeth M. Bilson, eds., *Cosmology and Astrophysics: Essays in Honor of Thomas Gold.* Ithaca, N.Y.: Cornell University Press, 1982.

Trefil, James S.:
The Moment of Creation. New York: Macmillan, 1983.
Space, Time, Infinity. New York: Pantheon Books, 1985.

Wagoner, Robert V., and Donald W. Goldsmith, *Cosmic Horizons.* San Francisco: W. H. Freeman, 1983.

Wald, Robert M., *Space, Time, and Gravity.* Chicago: University of Chicago Press, 1977.

Weinberg, Steven, *The First Three Minutes.* New York: Basic Books, 1977.

Westfall, Richard S., *Never at Rest.* Cambridge: Cambridge University Press, 1982.

Zeilik, Michael, and Elske v. P. Smith, *Introductory Astronomy and Astrophysics.* Philadelphia: Saunders College Publishing, 1987.

Periodicals

Abell, George O., "Cosmology—The Origin and Evolution of the Universe." *Mercury,* May/June 1978.

Alpher, Ralph A., Hans Bethe, and George Gamow, "The Origin of Chemical Elements." *Physical Review,* April 1, 1948.

Alpher, Ralph A., and Robert C. Herman, "Theory of the Origin and Relative Abundance Distribution of the Elements." *Reviews of Modern Physics,* April 1950.

Andersen, Per H., "Ripples in the Universal Hubble Flow." *Physics Today,* October 1987.

Bartusiak, Marcia:
"Before the Big Bang: The Big Foam." *Discover,* September 1987.
"If You Like Black Holes, You'll Love Cosmic Strings." *Discover,* April 1988.

"Bubbles upon the River of Time." *Science,* February 26, 1982.

Burbidge, E. Margaret, et al., "Synthesis of the Elements in Stars." *Reviews of Modern Physics,* October 1957.

Burns, Jack O., "Very Large Structures in the Universe." *Scientific American,* May 1986.

Corwin, Mike, and Dale Wachowiak, "Discovering the Expanding Universe." *Astronomy,* February 1985.

"Cosmic Cartography." *Scientific American,* March 1986.

"Cosmic Complex." *Scientific American,* January 1988.

Darling, David J.:
"Deep Time: The Fate of the Universe." *Astronomy,* January 1986.
"Universe 0:00." *Astronomy,* November 1980.

Davies, Paul:
"The Eleventh Dimension." *Science Digest,* January 1984.
"New Physics and the New Big Bang." *Sky and Telescope,* November 1985.
"Particle Physics for Everybody." *Sky and Telescope,* December 1987.
"Relics of Creation." *Sky and Telescope,* February 1985.

DeWitt, Bryce S., "Quantum Gravity." *Scientific American,* December 1983.

Dicke, R. H., et al., "Cosmic Black-Body Radiation." *Astrophysical Journal,* July 1, 1965.

Dressler, Alan, "The Large-Scale Streaming of Galaxies." *Scientific American,* September 1987.

Earman, John, and Clark Glymour, "Relativity and Eclipses: The British Eclipse Expeditions of 1919 and Their Predecessors." *Historical Studies in the Physical Sciences,* Vol. 2, Part 1, 1980.

Ferris, Timothy, "Where Are We Going?" *Sky and Telescope,* May 1987.

Fowler, William A., "The Quest for the Origin of the Elements." *Science,* November 1984.

Gaillard, Mary K., "Toward a Unified Picture of Elementary Particle Interactions." *American Scientist,* September/October 1982.

Gale, George, "The Anthropic Principle." *Scientific American,* December 1981.

Gamow, George:
"The Evolutionary Universe." *Scientific American,* September 1956.
"Expanding Universe and the Origin of Elements" (letter to the editor dated September 13, 1946). *Physical Review,* Vol. 70, page 572.
"The Origin of Elements and the Separation of Galaxies" (letter to the editor dated June 21, 1948). *Physical Review,* Vol. 74, page 505.

Georgi, Howard, "A Unified Theory of Elementary Particles and Forces." *Scientific American,* April 1981.

Gore, Rick, "The Once and Future Universe." *National Geographic,* June 1983.

Green, Michael B., "Superstrings." *Scientific American,* September 1986.

Gregory, Stephen A., "Our Cosmic Horizons—Part Three: The Structure of the Visible Universe." *Astronomy,* April 1988.

Guth, Alan, "The Eureka Moment." *Science Digest,* July 1985.

Guth, Alan, and Paul J. Steinhardt, "The Inflationary Universe." *Scientific American,* May 1984.

Harvey, Jeffrey A., "Superstrings." *Physics Today,* January 1987.

Hoyle, Fred:
"The Steady-State Universe." *Scientific American,* September 1956.
"The Synthesis of the Elements from Hydrogen." *Monthly Notices of the Royal Astronomical Society,* Vol. 106, 1946.

Kleist, T., "Lumps, Clumps and Jumps in the Universe." *Science News,* July 5, 1986.

Kragh, Helge, "The Beginning of the World: Georges Lemaître and the Expanding Universe." *Centaurus,* Vol.

32, 1987, pages 114-139.

Krauss, Lawrence M., "Dark Matter in the Universe." *Scientific American,* December 1986.

"Largest Supercluster Found." *Astronomy,* October 1985.

Lawrence, John K., "The Future History of the Universe." *Mercury,* November/December 1978.

Lemaître, Georges, "Supplement to 'Nature,' October 24, 1931." *Nature,* Vol. 128, 1931, pages 704-706.

Lemonick, Michael D., "The Universe: How Did It Begin?" *Science Digest,* October 1985.

Linde, Andrei, "Particle Physics and Inflationary Cosmology." *Physics Today,* September 1987.

LoPresto, James Charles, "The Geometry of Space and Time." *Astronomy,* October 1987.

MacRobert, Alan, "No Missing Mass?" *Sky and Telescope,* July 1985.

Mann, Charles C., and Robert P. Crease, "Waiting for Decay." *Science 86,* March 1986.

Melott, Adrian L., "Our Cosmic Horizons, Part Four: Re-creating the Universe." *Astronomy,* May 1988.

"Missing Neutrinos and Missing Mass." *Sky and Telescope,* November 1985.

Murphy, Jamie, "Bubbles in the Universe." *Time,* January 20, 1986.

Nambu, Yoichiro, "The Confinement of Quarks." *Scientific American,* November 1976.

"New Issues in Cosmology." *Physics Today,* January 1987.

"Nobel Prizes: Emphasis on Applications." *Science News,* October 20, 1979.

"The Nobel Prizes: That Winning American Style." *Time,* October 29, 1979.

Odenwald, Sten:
"The Decay of the False Vacuum." *Astronomy,* November 1983.
"Does Space Have More Than 3 Dimensions?" *Astronomy,* November 1984.
"The Planck Era." *Astronomy,* March 1984.
"To the Big Bang and Beyond." *Astronomy,* May 1987.

Overbye, Dennis, "The Shadow Universe." *Discover,* May 1985.

Penzias, A. A., and R. W. Wilson, "A Measurement of Excess Antenna Temperature at 4080 Mc/s." *Astrophysical Journal,* July 1, 1965.

"Physics: Unifying the Fields." *Newsweek,* October 29, 1979.

Quigg, C., "Elementary Particles and Forces." *Scientific American,* April 1985.

Rothman, Tony, and George Ellis, "Has Cosmology Become Metaphysical?" *Astronomy,* February 1987.

Rubin, Vera C., "Dark Matter in Spiral Galaxies." *Scientific American,* June 1983.

Sakharov, Andrei, "A Man of Universal Interests." *Nature,* February 25, 1988.

Schechter, Bruce, "Gravitation, Cosmology, and Cosmic-Ray Physics." *Physics Today,* April 1985.

Scherrer, Robert, "Our Cosmic Horizons, Part One: From the Cradle of Creation." *Astronomy,* February 1988.

Schramm, David, and Gary Stigman, "Particle Accelerators Test Cosmological Theory." *Scientific American,* June 1980.

Schwarzschild, Bertram, "Redshift Surveys of Galaxies Find a Bubbly Universe." *Physics Today,* May 1986.

"Searching for Primordial Pancakes." *Sky and Telescope,* November 1986.

Shankland, R. S., "The Michelson-Morley Experiment." *Scientific American,* November 1964.

Shu, Frank, "The Expanding Universe and the Large-Scale Geometry of Spacetime." *Mercury,* November/December 1983.

Silk, Joseph, Alexander S. Szalay, and Yakov B. Zel'dovich, "The Large-Scale Structure of the Universe." *Scientific American,* October 1983.

"Soapsuds Universe." *Sky and Telescope,* March 1986.

"Sponge Universe?" *Sky and Telescope,* December 1986.

Starobinskii, A. A., and Yakov B. Zel'dovich, "Quantum Effects in Cosmology." *Nature,* February 25, 1988.

"Super Collider Approved." *Sky and Telescope,* June 1987.

Taubes, Gary, "Everything's Now Tied to Strings." *Discover,* November 1986.

Thomsen, Dietrick E.:
"The Bottom Will Not Fall Out." *Science News,* May 17, 1986.
"Cosmic Caldron Bubbles Up Universe." *Science News,* February 20, 1982.
"In the Beginning Was Quantum Mechanics." *Science News,* May 30, 1987.
"The Quantum Universe: A Zero-Point Fluctuation?" *Science News,* August 3, 1985.
"Shadow Matter." *Science News,* May 11, 1985.

T'Hooft, Gerard, "Gauge Theories of the Forces between Elementary Particles." *Scientific American,* June 1980.

Tierney, John, "Exploding Star Contains Atoms of Elvis Presley's Brain." *Discover,* July 1987.

Trefil, James S.:
"The Accidental Universe." *Science Digest,* June 1984.
"How the Universe Will End." *Smithsonian,* June 1983.
" 'Nothing' May Turn Out to Be the Key to the Universe." *Smithsonian,* December 1981.

Trimble, Virginia, "Our Cosmic Horizons, Part Two: The Search for Dark Matter." *Astronomy,* March 1988.

Vilenkin, Alexander, "Cosmic Strings." *Scientific American,* December 1987.

Wagoner, Robert, and Donald Goldsmith, "Quarks, Leptons, and Bosons: A Particle Physics Primer." *Mercury,* July/August 1983.

Waldrop, M. Mitchell:
"Before the Beginning." *Science 84,* January/February 1984.
"Do-It-Yourself Universes." *Research News,* February 20, 1987.
"The Large-Scale Structure of the Universe Gets Larger—Maybe." *Science,* November 13, 1987.
"Supersymmetry and Supergravity." *Science,* April 29, 1983.

INDEX

Mass: average density of, in universe, consequences of, 37-38, *56-57*, 101-102, 118; in Einstein's equation, 30, 46; electron volts for expressing, 95; and gravitational force, in Newton's theory, 19; increase of, in special relativity, *46;* of neutrinos, question of, 111, 116; of quarks, 87; unseen (dark matter), 116-118
Matrix approach, Heisenberg's, 68
Matter: dark, 116-118; excess of, over antimatter, 123, 126; Steady State creation of, 77, 78, 79; ylem, 75
Matter Era, 76, *130-131*
Maxwell, James Clerk, 27, 64, 66
Maxwell's equations, 27, 29
Mercury (planet), perihelion of, *33*
Michelson, Albert, experiment by, *26, 27*
Microwave radiation, background, 79-81, 82, 100
Milky Way galaxy, *17*
Minkowski, Hermann, diagraming technique of, *47*
Moon, orbital period of, Newton's calculation of, 19, 24
Morley, Edward, experiment by, *26, 27*
Motion: Newton's laws of, 18, 26. *See also* Acceleration; Special relativity
Muons, *100;* moving, extended time scales of, 42
Musset, Paul, 97

N

Nebulae, extragalactic (galaxies), 61
Neptune (planet), position of, 26
Neutral currents, search for, 96-97
Neutrino Era, 127
Neutrinos, *84,* 96-97, *124, 127;* detecting, 111, 116, *117;* free-streaming, Zel'dovich's, 110, 111; mass, question of, 111, 116; protons disintegrated by, *58-59,* 111
Neutrons, 75, *117;* in atom formation, *131;* in Nucleosynthesis Era, *128;* quarks forming, 86
Newton, Isaac, 14-19, 24-26; and gravitation, theory of, 18-19, 24, 25-26, 32; light, theory of, *73;* and the *Principia,* 24-25; quoted, *17,* 19, 24, 25
Night sky, reasons for darkness of, *20-23*
Non-Euclidean geometry, 32, 36, 55, *57*
Nuclei, atomic, 71; in atom formation, *130-131;* bombarded with protons, 74; formation of, *128-129;* in Rutherford model, 66

Nucleon-neutrino interactions, 96-97
Nucleosynthesis Era, *128-129*
Nucleosynthesis within stars, 79
Nuffield Workshop on the Very Early Universe, 109
Numbering system for cosmology, 99

O

Oort, Jan Hendrik, 117
Open universe, 37, 56, *57*
Orbits: of electrons, concept of, 67, 68; of Mercury, *33;* of Moon, period of, Newton's calculation of, 19, 24

P

Particle accelerators, 74, 92, 95, 97
Particles, 12, *58-59,* 83, *84-89,* 92-93, *120;* alpha, 65, 66, 74; in chambers, *4-5, 100-101;* in Electroweak Era, *124-125;* in GUT Era, *120-121;* in Inflation Era, *122-123;* light as, 64, *73;* mass of, expressing, 95; muons, 42, *100;* neutrino detection, 111, 116, *117;* in Nucleosynthesis Era, *128-129;* in Quark Confinement Era, *126-127;* uncertainty principle and, 69, *70-71, 88, 89;* and unification theories, 94-95, 96-97, 98-100; virtual, 85, *88-89;* as waves, 68-69; ylem, 75. *See also* Atoms
Peebles, James, 81, 101-102
Penzias, Arno, 81, 82; antenna used by, *80,* 82; Gamow's note to, *81*
Perfect cosmological principle, 78
Phase transitions, 102, 103, 108, 112
Photoelectric effect, 64
Photons, *85,* 94, *120;* atom formation allowed by weakening of, *130-131;* Einstein's theory of, 64; in Electroweak Era, *124, 125;* in Nucleosynthesis Era, 128, *129;* in Quark Confinement Era, *126-127;* uncertainty principle and, *70-71, 88;* virtual, *88, 89*
Physical Review, Gamow-Alpher paper in, 75
Picard, Jean, 19
Pions, *100*
Planck, Max, 30, *63*-64, 65
Planck Era, *104-105*
Planck length, 104-*105*
Planck's constant, 64; Bohr's use of, 67
Planck Time, 105
Poe, Edgar Allan, quoted, 20
Positrons, *84, 89;* in Electroweak Era, *124, 125;* in Quark Confinement Era, *126*
Primeval atom, Lemaître's, 61, 62, 74
Principia (The Mathematical Principles of Natural Philosophy; New-

ton), 24-25
Protons, *127;* and antiprotons, *100-101, 127;* in atom formation, *131;* decay of, spontaneous, predicted, 99-100; disintegration of, by impact, *58-59,* 111, *117;* formed from quarks, 86, *87,* 126; nuclei bombarded with, 74; in Nucleosynthesis Era, *128, 129*

Q

Quanta of energy, 63, 64-65, 67
Quantum chromodynamics (QCD), 98
Quantum electrodynamics (QED), 93
Quantum fluctuations and galaxy formation, 109
Quantum gravity, 105
Quantum mechanics, 63, 92-93. *See also* Atoms
Quark Confinement Era, *126-127*
Quarks, *84, 120, 121;* confinement of, *126;* in Electroweak Era, *125;* exchange of gluons between, *86;* in Inflation Era, *123;* protons and neutrons formed from, 86, *87,* 126; types, properties of, 87 *chart*

R

Radiation: background, cosmic, 79-81, 82, 100; black body, 63-64
Radio antenna, *80,* 82
Red and green quarks, *86*
Red shifts: cosmological, *22-23,* 77; Doppler, 22, *23*
Relativity, 39; and quantum mechanics, 70-71. *See also* General relativity; Special relativity
Renormalization, 93; problems of, in electroweak unification, 94-95, 96
Repulsion theory for strings, *114-115*
Rest vs. motion, 26
Riemann, Bernhard, 32
Roll, Peter, 81
Royal Astronomical Society, 34, 62
Royal Society of London, 19
Rubin, Vera, 117
Rutherford, Ernest, *65*-66; and Gamow, 74; quoted, 66, 68

S

Salam, Abdus, 95-96, *97,* 98
Schrödinger, Erwin, 68-69
Schuster, Arthur, 65
Schwinger, Julian, 93-94
Scientific notation, 99
Simultaneity, loss of, in special relativity, 40, *41*
Solar eclipses, expeditions to observe, 33, 35, 53
Space-time: continuum of, in special relativity, *47;* topography of, in infant universe, *105. See also* Curvature of space-time

ACKNOWLEDGMENTS

The editors wish to thank Ralph Alpher, Union College, Schenectady, N.Y.; Reinhard Bachmann, Rudolf Heinrich, Deutsches Museum, Munich; E. Benamy, American Friends of Hebrew University, New York; Pierre Berthon, Académie des Sciences, Institut de France, Paris; Thomas H. Callen, National Air and Space Museum, Washington, D.C.; Michelangelo De Maria, Università degli Studi, Rome; S. Djorgovski, California Institute of Technology, Pasadena; John C. Evans, George Mason University, Fairfax, Va.; Igor Gamow, University of Colorado, Boulder; Ann Gatti, International Center for Theoretical Physics, Trieste, Italy; Margaret Geller, Harvard Smithsonian Center for Astrophysics, Cambridge, Mass.; Sheldon Glashow, Brookline, Mass.; Judith Goldhaber, Lawrence Berkeley Laboratory, Berkeley, Calif.; Alan H. Guth, Massachusetts Institute of Technology, Cambridge; Joe Harris, Dartmouth College, Hanover, N.H.; John Heilbron, Berkeley, Calif.; Peter D. Hingley, Librarian, Royal Astronomical Society, London; Charles Hurley, Lawrence Berkeley Laboratory, Berkeley, Calif.; Christine Jones, Harvard Smithsonian Center for Astrophysics, Cambridge, Mass.; Ann Kottner, American Institute of Physics, New York; Helge Kraghe, Cornell University, Ithaca, N.Y.; Anthony Longhitano, Middletown, N.J.; David Malin, Anglo-Australian Observatory, Sydney; Philip Menza, Black Star, New York; Messieurs les Secrétaires Perpetuels, Académie des Sciences, Institut de France, Paris; Giuseppe Monaco, Curator, Museo Astronomico e Copernicano, Rome; Alan Needell, National Air and Space Museum, Washington, D.C.; Helmut Rechenberg, Max Planck Institut für Physik und Astrophysik, Munich; Frederick Reines, University of California, Irvine; Abdus Salam, International Center for Theoretical Physics, Trieste, Italy; Paul J. Steinhardt, University of Pennsylvania, Philadelphia; Philip Taylor, Case Western Reserve University, Cleveland, Ohio; Alexander Vilenkin, Tufts University, Medford, Mass.

PICTURE CREDITS

The sources for the illustrations that appear in this book are shown below. Credits from left to right are separated by semicolons, from top to bottom by dashes.

Cover: Art by Paul Hudson. Front and back endpapers: Computer-generated art by John Drummond. 2, 3: European Southern Observatory, Garching, FRG. 4, 5: Courtesy Lawrence Berkeley Laboratory. 10, 11: Michel Tcherevkoff. 12: Initial cap, detail from pages 2, 3. 15: Ann Ronan Picture Library, Taunton, Somerset, England. 16: Bibliothèque Nationale, Paris; courtesy Explorer Archives. 17: Mount Wilson and Las Companas Observatories, Carnegie Institution of Washington (4)—Hale Observatories/Palomar Observatory, © California Institute of Technology. 20-23: Art by Damon M. Hertig and Dan Rodriguez, the Art Connection. 26: Art by Fred Holz. 29: Hebrew University of Jerusalem, courtesy AIP Niels Bohr Library. 33: Art by Damon M. Hertig and Dan Rodriguez, the Art Connection. 34: Winifred Eddington, copied by Larry Sherer, from *The Life of Arthur Stanley Eddington*, by A. Vibert Douglas, Thomas Nelson and Sons Ltd., 1956. 35: Art by Damon M. Hertig and Dan Rodriguez, the Art Connection. 37: Courtesy AIP Niels Bohr Library. 39-57: Art by Damon M. Hertig and Dan Rodriguez, the Art Connection. 58, 59: Patrice Loiz, CERN/Science Photo Library. 60: Initial cap, detail from pages 58, 59. 61: Archive de L'Académie Royale de Belgique, No. 15461, photo Luc Schrobiltgen. 63: Courtesy AIP Niels Bohr Library, Lande Collection. 65: L. E. Wynn-William/Niels Bohr Library. 67: Courtesy Niels Bohr Library, Uhlenbeck Collection. 69: Culver Pictures. 70, 71: Art by Damon M. Hertig and Dan Rodriguez, the Art Connection. 72: AIP Niels Bohr Library, from negatives provided by Martin J. Klein, restored by William R. Whipple. 73: Art by Fred Holz. 75: Prof. Anthony J. G. Hey, Southampton, England. 76: Carl Iwasaki, courtesy Elfrieda and Igor Gamow, Denver, Colorado. 77: Tom Nebbia, courtesy National Geographic Society. 80: Bell Telephone Laboratories, courtesy National Geographic Society. 81: Reproduced with permission of AT & T Archives. 83-89: Art by Yvonne Gensurowsky of Stansbury, Ronsaville, and Wood, Inc. 90-91: James Sugar/Black Star, computer image by S. Djorgovski. 92: Initial cap, detail from pages 90, 91. 100: Lawrence Berkeley Laboratory/Science Photo Library; Prof. G. Piragino/Science Photo Library. 101: CERN (Conseil Europeen pour la Recherche Nucléaire). 103: Henry R. Hilliard, courtesy Dr. Alan Guth. 104-107: Art by Stephen Wagner. 111: Courtesy Smithsonian Astrophysical Observatory. 112-117: Art by Stephen Wagner. 119: Art by Yvonne Gensurowsky of Stansbury, Ronsaville, and Wood, Inc. 120, 121: Art by Alfred Kamajian and Yvonne Gensurowsky. 122-131: Art by Alfred Kamajian.

Time-Life Books Inc.
is a wholly owned subsidiary of
TIME INCORPORATED

FOUNDER: Henry R. Luce 1898-1967

Editor-in-Chief: Jason McManus
Chairman and Chief Executive Officer:
J. Richard Munro
President and Chief Operating Officer:
N. J. Nicholas, Jr.
Editorial Director: Ray Cave
Executive Vice President, Books: Kelso F. Sutton
Vice President, Books: Paul V. McLaughlin

TIME-LIFE BOOKS INC.
EDITOR: George Constable
Executive Editor: Ellen Phillips
Director of Design: Louis Klein
Director of Editorial Resources: Phyllis K. Wise
Editorial Board: Russell B. Adams, Jr., Dale M.
Brown, Roberta Conlan, Thomas H. Flaherty, Lee
Hassig, Donia Ann Steele, Rosalind Stubenberg
Director of Photography and Research:
John Conrad Weiser
Assistant Director of Editorial Resources:
Elise Ritter Gibson

PRESIDENT: Christopher T. Linen
Chief Operating Officer: John M. Fahey, Jr.
Senior Vice Presidents: Robert M. DeSena, James
L. Mercer, Paul R. Stewart
Vice Presidents: Stephen L. Bair, Ralph J. Cuomo,
Neal Goff, Stephen L. Goldstein, Juanita T.
James, Hallett Johnson III, Carol Kaplan, Susan
J. Maruyama, Robert H. Smith, Joseph J. Ward
Director of Production Services: Robert J.
Passantino
Supervisor of Quality Control: James King

Editorial Operations
Copy Chief: Diane Ullius
Production: Celia Beattie
Library: Louise D. Forstall

Correspondents: Elisabeth Kraemer-Singh (Bonn),
Maria Vincenza Aloisi (Paris), Ann Natanson
(Rome). Valuable assistance was also provided by
Brigid Grauman (Brussels), Christine Hinze
(London), Christina Lieberman (New York), Ann
Wise (Rome), Mary Johnson (Stockholm), Traudl
Lessing (Vienna).

VOYAGE THROUGH THE UNIVERSE

SERIES DIRECTOR: Roberta Conlan
Series Administrator: Judith W. Shanks

Editorial Staff for *The Cosmos*
Designer: Dale Pollekoff
Associate Editor: Sally Collins (pictures)
Text Editors: Peter Pocock (principal), Pat Daniels
Researchers: Tina S. McDowell, Barbara C.
Mallen, Barbara Sause
Writer: Esther Ferington
Assistant Designer: Brook Mowrey
Editorial Assistant: Alice T. Marcellus
Copy Coordinator: Darcie Conner Johnston
Picture Coordinator: Richard Karno

Special Contributors: Linda Billings, William
Cromie, Peter Gwynne, Neil McAleer, Gina
Maranto, John I. Merritt, Wallace Tucker, Mark
Washburn (text); Lynn Cook, Ron Cowen, Kirk
Denkler, Marilyn Fenischel, Sanjoy Ghosh,
Jocelyn Lindsay, Shawn McCarthy, Hugh Mc-
Intosh, Jacqueline Shaffer, Cindy Spitzer, Julie
Ann Trudeau (research); Barbara L. Klein (index).

CONSULTANTS

GEOFFREY BURBIDGE is a professor of physics at
the University of California, San Diego. Formerly
the director of Kitt Peak National Observatory, he is
known for his work on the origin of cosmic rays,
extragalactic astronomy, and stellar nucleosynthesis.

MINAS KAFATOS, a theoretical astrophysicist,
teaches cosmology at George Mason University and
quantum theory at the Smithsonian Institution.

DELO E. MOOK teaches physics and astronomy at
Dartmouth College. He is also collaborating on a
comparison of the development of modern physics
with that of twentieth-century art and literature.

STEN ODENWALD, principal consultant for this
volume, is an infrared astronomer at the Naval Research Laboratory. He also teaches at the Smithsonian Institution and has published widely on
cosmology.

ROBERT W. SMITH is a historian of astronomy who
holds joint appointments at the National Air and
Space Museum, Smithsonian Institution, and in the
History of Science Department of the Johns Hopkins
University.

JAMES TREFIL, a professor of physics at George
Mason University, has won several awards for his
publications for nonscientists.

**Library of Congress Cataloging in
Publication Data**
The Cosmos/by the editors of Time-Life Books.
p. cm. (Voyage through the universe).
Bibliography: p.
Includes index.
ISBN 0-8094-6862-X.
ISBN 0-8094-6863-8 (lib. bdg.).
1. Cosmology. I. Time-Life Books. II. Series.
QB981.C826 1989
523.1—dc19   88-24945 CIP

For information on and a full description of any
of the Time-Life Books series, please call 1-800-
621-7026 or write:
Reader Information
Time-Life Customer Service
P.O. Box C-32068
Richmond, Virginia 23261-2068

Time-Life Books Inc. offers a wide range of fine
recordings, including a *Rock 'n' Roll Era* series.
For subscription information, call 1-800-621-7026
or write Time-Life Music, P.O. Box C-32068,
Richmond, Virginia 23261-2068.

Earth: diameter 7,926 miles

Neptune: diameter 30,200 miles

Uranus: diameter 31,600 miles

Red supergiant: diameter 400 million miles

Solar System: diameter 7.5 billion miles

Globular cluster: diameter 2×10^{14} miles

Milky Way: diameter 100,000 light-years

Local Group of galaxies:
6 million light-years across

Largest double radio source:
length 17 million light-years